CAD in Machinery Design: Implementation and Educational Issues (CADMD 2020)

XXVIII International Ukrainian-Polish Scientific and Technical Conference

IOP Conference Series: Materials Science and Engineering
Volume 1016

Lviv, Ukraine
26 - 27 November 2020

Editors:

Mykhailo Lobur
Nazariy Jaworski

ISBN: 978-1-7138-3306-2
ISSN: 1757-8981

Printed from e-media with permission by:

Curran Associates, Inc.
57 Morehouse Lane
Red Hook, NY 12571

Some format issues inherent in the e-media version may also appear in this print version.

Institute of Physics
Dirac House, Temple Back
Bristol BS1 6BE UK

Phone: 44 1 17 929 7481
Fax: 44 1 17 920 0979

techtracking@iop.org

Additional copies of this publication are available from:

Curran Associates, Inc.
57 Morehouse Lane
Red Hook, NY 12571 USA
Phone: 845-758-0400
Fax: 845-758-2633
Email: curran@proceedings.com
Web: www.proceedings.com

TABLE OF CONTENTS

IOP Conf. Series: Materials Science and Engineering 1016 (2021) 011001 doi:10.1088/1757-899X/1016/1/011001

CADMD 2020

Conference Proceedings

XXVIII International Ukrainian-Polish
Scientific and Technical Conference

CAD IN MACHINERY DESIGN: IMPLEMENTATION AND EDUCATIONAL ISSUES

26-27 November 2020, Lviv, Ukraine

http://cadmd.lpnu.ua

cadmd@lpnu.ua

In connection with the quarantine restrictions caused by the spread of coronavirus infection 2019-nCoV in Ukraine and around the world the CADMD 2020 conference was held in online mode

ORGANIZED BY

**Department of Computer
Aided Design Systems,
Lviv Polytechnic National University**

**Faculty of Electrical Engineering,
Bialystok University of Technology**

**Mechanical Engineering and Robotics,
AGH University of Science and Technology**

**Institute of Machine Design Fundamentals,
Warsaw University of Technology**

OFFICIAL PARTNER

KSSE
Katowice
Special
Economic Zone

www.ksse.com.pl
ksse@ksse.com.pl
+48 32-251-07-36

IOP Conf. Series: Materials Science and Engineering 1016 (2021) 011001 doi:10.1088/1757-899X/1016/1/011001

Preface

The CADMD conference series was started in Warsaw in 1994 and has been organized yearly in Poland and Ukraine.

The goal of the conference is to exchange experience between scientists of Ukraine, Poland and their young, promising colleagues in the field of modern information technologies, development of CAD in mechanics, methodology of teaching and implementation of automated systems in the educational process and to establish closer links between researchers.

These proceedings contain 23 presented at the CADMD 2020 conference papers in such topics on CAD in machinery design as:

- information technologies in computer-aided design;
- modeling and simulation of processes and systems in machinery design;
- microelectromechanical systems design, simulation, modeling and application;
- CAD systems in mechatronics, robotics, and automation engineering;
- CAD tools in industry 4.0.;
- engineering Education.

The papers were contributed by 71 authors from such institutions as:

- Ukrainian National Forestry University;
- Lviv Polytechnic National University;
- Poznan University of Technology;
- Warsaw University of Technology;
- Taras Shevchenko National University of Kyiv;
- AGH University of Science and Technology;
- Karpenko Physico-Mechanical Institute of the National Academy of Sciences of Ukraine;
- National Technical University Dnipro Polytechnic;
- Dnipro State Agrarian and Economic University;
- Danylo Halytsky Lviv National Medical University;
- National University Zaporizhzhia Polytechnic;
- Kazimierz Wielki University in Bydgoszcz;
- Bialystok University of Technology;
- Kharkiv Polytechnic Institute;
- PGNiG Termika Energetyka Przemysłowa.

Conference Proceedings Editors

prof. Mykhailo Lobur, conference co-chairman, head of Computer-Aided Design Systems Department, Lviv Polytechnic National University, 79013, Mytropolyta Andreya St. 3, Lviv, Ukraine, mykhaylo.v.lobur@lpnu.ua;

dr. Nazariy Jaworski, conference secretary, assistant professor of Computer-Aided Design Systems Department, Lviv Polytechnic National University, 79013, Mytropolyta Andreya St. 3, Lviv, Ukraine, nazarii.b.yavorskyi@lpnu.ua.

Peer Review Statement

All papers published in this volume of *IOP Conference Series: Materials Science and Engineering* have been peer reviewed through processes administered by the Editors. Reviews were conducted by expert referees to the professional and scientific standards expected of a proceedings journal published by IOP Publishing."

IOP Conf. Series: Materials Science and Engineering 1016 (2021) 011001 doi:10.1088/1757-899X/1016/1/011001

Disclaimer

This year organizers wanted to continue the conference tradition and invite participants to one of the most attractive city in Ukraine – Lviv, but due to the rapid worldwide spreading of the coronavirus infection (COVID-2019) and announced quarantine in Ukraine and many other EU countries, the CADMD 2020 Organizing Committee has decided to run the conference in a distance on-line form. COVID-19 and quarantine restrictions should not stop the exchange of an experience between scientists of Ukraine, Poland and their young, promising colleagues in the field of modern information technologies, development of CAD in mechanics, methodology of teaching and implementation of automated systems in the educational process and to establish closer links between researchers.

This is a new challenge for all of us, and we appreciate participants who have submitted their papers. Each member of the Organization Committee did his best to make the Conference interesting and useful for each participant in order to strengthen our research collaboration and scientific partnership.

The CADMD 2020 conference was held on 26-27 November 2020 online basing on the Cisco Webex Video Conferencing system. Recorded streaming is available at the conference web site http://cadmd.lpnu.ua.

The conference was consisting of:

- Opening session – 20 min;
- Session A – 70 min, 4 presenters;
- Session B – 90 min, 7 presenters;
- Session C – 90 min, 7 presenters;
- Session D – 90 min, 6 presenters;
- Closing session – 15 min;

Each paper presentation has taken up to 10 min, discussion and questions – up to 5 min. The audience was consisting of near 50 unique attenders form Poland and Ukraine.

CADMD 2020 Conference Organizing Committee,
Department of Computer Aided Design Systems,
Lviv Polytechnic National University

Education building IV, room 320,
79013, Mytropolyta Andreya St. 3, Lviv, Ukraine.
Tel.: +38 032 258 26 74
Email: cadmd@lpnu.ua
Internet: http://cadmd.lpnu.ua

IOP Conf. Series: Materials Science and Engineering 1016 (2021) 011001 doi:10.1088/1757-899X/1016/1/011001

International Organizing Committee

Chairmen:

prof. Mykhailo Lobur	Lviv Polytechnic National University, Ukraine
prof. Roman Kaczyński	Technical University of Białystok, Poland
prof. Robert Zalewski	Warsaw University of Technology, Poland
prof. Marian Banaś	AGH University of Science and Technology

Conference Secretary:

dr. Nazariy Jaworski	Lviv Polytechnic National University, Ukraine

Members of Organizing Committee:

dr. Andriy Kernytskyy	Lviv Polytechnic National University, Ukraine
dr. Mykhaylo Melnyk	
dr. Uliana Marikutsa	
dr. Ihor Farmaha	
dr. Oksana Oborska	
msc. Volodymyr Dutka	
msc. Bohdan Vasylyshyn	

International Programme Committee

prof. Aivars Aboltins	Latvia University of Life Sciences and Technologies
prof. Jerzy Bajkowski	Warsaw University of Technology
prof. Marian Banaś	AGH University of Science and Technology
prof. Bogusław Butryło	Białystok University of Technology
prof. Marek Iwaniec	AGH University of Science and Technology
dr. Nazariy Jaworski	Lviv Polytechnic National University
prof. Roman Kaczyński	Białystok University of Technology
prof. Antoni Kalukiewicz	AGH University of Science and Technology
prof. Krzysztof Kołodziejczyk	AGH University of Science and Technology
prof. Petro Kosobutskyy	Lviv Polytechnic National University
prof. Krzysztof Krawiec	Poznań University of Technology
prof. Mykhailo Lobur	Lviv Polytechnic National University
dr. Andrzej Łukaszewicz	Białystok University of Technology
prof. Oleh Matviykiv	Lviv Polytechnic National University
dr. Mykhaylo Melnyk	Lviv Polytechnic National University
dr. Grzegorz Mieczkowski	Białystok University of Technology
prof. Jerzy Pokojski	Warsaw University of Technology
prof. Krzysztof Pytel	AGH University of Science and Technology
prof. Anna Timofiejczuk	Silesian University of Technology
prof. Wojciech Sitek	Silesian University of Technology
dr. Roman Sheremeta	Lviv Polytechnic National University
prof. Yaroslav Sokolovskyy	Lviv Polytechnic National University
prof. Adam Sołbut	Białystok University of Technology
prof. Maciej Zajkowski	Białystok University of Technology
prof. Robert Zalewski	Warsaw University of Technology

IOP Conf. Series: Materials Science and Engineering 1016 (2021) 011002 doi:10.1088/1757-899X/1016/1/011002

Peer review declaration

All conference organisers/editors are required to declare details about their peer review. Therefore, please provide the following information:

- **Type of peer review: Single-blind / Double-blind / Triple-blind / Open / Other (please describe)**
  Open - each submitted paper was evaluated by reviewers, members of the Programme Committee (2-4 reviewers per submission depending on decisions about rejecting or accepting after revision). Their remarks and suggestions were sent back to authors by email in order to facilitate timely preparation of revised papers.
- **Conference submission management system:**
  Gmail, Google Drive
- **Number of submissions received:**
  34
- **Number of submissions sent for review:**
  34
- **Number of submissions accepted:**
  23
- **Acceptance Rate (Number of Submissions Accepted / Number of Submissions Received X 100):**
  67%
- **Average number of reviews per paper:**
  3
- **Total number of reviewers involved:**
  14
- **Any additional info on review process:**
  All accepted submissions have passed 30% barrier of similarity index in the IEEE Cross Check system
- **Contact person for queries:**
  CADMD conference secretary,
  assistant professor of the CAD Systems Department
  of Lviv Polytechnic National University
  Ph.D. Nazariy Jaworski
  cadmd@lpnu.ua

IOP Conf. Series: Materials Science and Engineering 1016 (2021) 012001 doi:10.1088/1757-899X/1016/1/012001

Mathematical Modeling of Boundary Stress State of Orthotropic Material

Sofia Holovata[1], Yaroslav Sokolovskyy[1], Bohdan Pobereyko[2] and Andriy Holovatyy[3]*

[1]Department of Information Technologies, Ukrainian National Forestry University (UNFU), 103 Gen. Chuprynka St., 79057 Lviv, Ukraine
[2]Department of Automation and Computer-Integrated Technologies, Ukrainian National Forestry University (UNFU), 11 Zaliznyaka St., 79044 Lviv, Ukraine
[3]Department of Computer-Aided Design Systems, Lviv Polytechnic National University, 5 Mytropolyta Andreya St., building 4, 79013 Lviv, Ukraine

*Email: andrii.i.holovatyi@lpnu.ua

Abstract. In the paper, the mathematical model for determination of stress and deformation fields inside and at the elastic region boundary of deformation of orthotropic materials under the biaxial loadings is developed. Within the proposed model, a mathematical model for determination of the material transition boundary from elastic into non-elastic deformation region is developed and a system of nonlinear equations for calculation of the values of its parameters is obtained. By using the continuous method of solution continuation with respect to the best parameter and Runge-Kutta method, the demarcation curves of absolutely elastic and non-elastic deformation regions for the pine tree are constructed. The analysis of the obtained curves is carried out and on its base it is shown that the proposed model, in contrast to the known mechanical theories of strength, makes possible to determine and substantiate the ultimate stress states of materials whose asymmetry coefficients of the boundaries of proportionality in some anisotropy directions are greater, and in others less than one.

1. Introduction

Nowadays, for modeling the boundary stress states of isotropic and anisotropic materials the implicit functions of two or more unknowns are generally used. These functions are determined by nonlinear algebraic equations of the second and higher orders. The equations with two unknowns are used to model biaxial boundary stress states in the planes of structural symmetry of the material and equations with three or more unknowns are used to solve the problem in the planar and more complex formulations. In the deformable solid mechanics, they are usually called criteria, or conditions of strength. For today, more than 40 such criteria have been proposed [1]. In particular, the criteria of Ashkenazi, Goldenblatt-Kopnov, Mises Hill, Zakharov and others belong to them.

However, their practical application is limited. They are mainly used only to approximate the results of experimental studies of the ultimate stress states of the mentioned materials [1].

IOP Conf. Series: Materials Science and Engineering 1016 (2021) 012001 doi:10.1088/1757-899X/1016/1/012001

These equations are not derived, they are phenomenological. They do not make possible to substantiate the peculiarities of deformability inside and on the boundary of the absolutely elastic region of deformation of solids with strong and weak asymmetries of strength characteristics [2, 3].

Therefore, the task of developing such mathematical model is actual, which, in contrast to existing phenomenological models, would make possible not only to approximate the data of empirical studies, but also to identify and justify the main factors of asymmetry characteristics of the ultimate stress state of the studied material.

The practical expediency of its solution is substantiated by the fact that today there are no models that would satisfactorily describe the ultimate stress states, the values of the asymmetry characteristics of which in some directions of structural symmetry of anisotropic materials are 2 ÷ 3 times larger, and in others - 2 ÷ 3 smaller per unit.

In addition, for today there are no technical solutions for the timely detection of the transition of these materials from absolutely elastic to non-elastic deformation region in the conditions of their processing and subsequent operation.

2. Mathematical model for determination of the absolutely elastic deformation region of orthotropic materials

In order to solve this problem, the mathematical model is synthesized [4], its components are:

- Assumption: in the extreme stress state the potential energy density of the material under biaxial loading) $(\sigma_{11} \neq 0; \sigma_{22} \neq 0; \sigma_{12} = \sigma_{13} = \sigma_{23} = 0)$ is directly proportional to the relative bulk deformations of the material tested in the conditions of the uniaxial stretching (compressions) in the orthotropic directions, i.e.:

$$\Pi(\sigma_{11};\sigma_{22}) = k_1\varepsilon_v(\sigma_{11};0) + k_2\varepsilon_v(0;\sigma_{22}) + k_3; \tag{1}$$

- The formula for determination of the potential energy density of the deformed material:

$$\Pi(\sigma_{11};\sigma_{22}) = 0{,}5\sum_{i=1}^{2}\sigma_{ii}\varepsilon_{ii}(\sigma_{11};\sigma_{22}) \tag{2}$$

- The formula for determining the relative bulk deformation:

$$\varepsilon_V(\sigma_{11};\sigma_{22}) = \sum_{i=1}^{3}\varepsilon_{ii}(\sigma_{11};\sigma_{22}); \tag{3}$$

- Hooke's law for compressible orthotropic materials:

$$\begin{cases} \varepsilon_{11} = L\left(\dfrac{1-\mu_{23}\mu_{32}}{E_{11}}\sigma_{11} + \dfrac{\mu_{13}\mu_{32}-\mu_{12}}{\sqrt{E_{11}E_{22}}}\sigma_{22}\right); \\ \varepsilon_{22} = L\left(\dfrac{\mu_{23}\mu_{31}-\mu_{21}}{\sqrt{E_{11}E_{22}}}\sigma_{11} + \dfrac{1-\mu_{13}\mu_{31}}{E_{22}}\sigma_{22}\right); \\ \varepsilon_{33} = L\left(\dfrac{\mu_{32}\mu_{21}-\mu_{31}}{\sqrt{E_{11}E_{33}}}\sigma_{11} + \dfrac{\mu_{31}\mu_{12}-\mu_{32}}{\sqrt{E_{22}E_{33}}}\sigma_{22}\right); \\ \varepsilon_{12} = \varepsilon_{13} = \varepsilon_{23} = 0. \end{cases} \tag{4}$$

Here:

$$L^{-1} = 1 + \mu_{12}\mu_{23}\mu_{31} + \mu_{21}\mu_{13}\mu_{32} - \mu_{13}\mu_{31} - \\ - \mu_{23}\mu_{32} - \mu_{12}\mu_{21}; \tag{5}$$

σ_{ii} - components of the stress tensor; $\varepsilon_{ii}(\sigma_{11};\sigma_{22})$ - dependences of the components of the strain tensor on the components of the stress tensor; $\Pi(\sigma_{11};\sigma_{22})$-dependence of the potential energy density of the deformed material on the components of the stress tensor; μ_{ij} – Poisson's ratios; E_{ii} – modules of elasticity; k_1, k_2 and k_3 – parameters depending on the signs of the stress tensor components σ_{ii}.

Within the framework of the proposed model (1) - (5) the mathematical model for determining the region boundary of the absolutely elastic behavior of the orthotropic material under the biaxial loading conditions:

$$\Phi(\sigma_{11},\sigma_{22})=0, \tag{6}$$

where

$$\Phi(\sigma_{11},\sigma_{22})=\frac{1-\mu_{23}\mu_{32}}{2E_{11}}\sigma_{11}^2+\left(\frac{\mu_{13}\mu_{32}+\mu_{23}\mu_{31}}{2\sqrt{E_{11}E_{22}}}-\right. \tag{7}$$

$$\left.-\frac{\mu_{12}+\mu_{21}}{2\sqrt{E_{11}E_{22}}}\right)\sigma_{11}\sigma_{22}+\frac{1-\mu_{13}\mu_{31}}{2E_{22}}\sigma_{22}^2+\frac{D}{L}\sigma_{11}+.$$

$$\frac{E}{L}\sigma_{22}+\frac{F}{L}.$$

The values of D, E and F are calculated from the following system of the equations:

$$\begin{cases}\varphi_1(D,E)=0;\\ \varphi_2(D,E)=0;\\ NF=-a_{11}D-a_{22}E-b,\end{cases} \tag{8}$$

where

$$\varphi_1(D,E)=\left(\frac{3b-2}{N}a_{11}-b_{11}\right)D^2+\left(\frac{b^2}{N}-c\right)D+ \tag{9}$$

$$+\left(\frac{a_{11}a_{22}}{N}-a_{12}\right)D^2E+\left(q+\frac{a_{22}^2-a_{11}^2}{N}\right)DE^2+2\cdot$$

$$\cdot\left(\frac{ba_{22}}{N}-b_{22}\right)DE+\left(b_{11}-\frac{ba_{11}}{N}\right)E^2+\left(a_{12}-\frac{a_{11}a_{22}}{N}\right)E^3$$

$$\varphi_2(D,E)=\left(\frac{ba_{22}}{N}-b_{22}\right)E^2+\left(\frac{b^2}{N}-c\right)E+ \tag{10}$$

$$+\left(\frac{a_{11}a_{22}}{N}-a_{12}\right)DE^2+\left(\frac{b^2}{N}-c\right)E+\left(\frac{a_{22}^2-a_{11}^2}{N}-q\right)\cdot$$

$$\cdot D^2E+2\left(\frac{a_{11}(2b-1)}{N}-b_{11}\right)DE+\left(b_{22}-\frac{ba_{22}}{N}\right)D^2+$$

$$+\left(a_{12}-\frac{a_{11}a_{22}}{N}\right)D^3$$

$$a_{11}=\sum_{n=1}^{N}\tilde{\sigma}_{11}^{(n)}; \qquad a_{22}=\sum_{n=1}^{N}\tilde{\sigma}_{22}^{(n)}; \qquad a_{12}=\sum_{n=1}^{N}\tilde{\sigma}_{11}^{(n)}\tilde{\sigma}_{22}^{(n)};$$

$$b=\sum_{n=1}^{N}\Pi\left(\tilde{\sigma}_{11}^{(n)};\tilde{\sigma}_{22}^{(n)}\right); \qquad c=\sum_{n=1}^{N}\Pi^2\left(\tilde{\sigma}_{11}^{(n)};\tilde{\sigma}_{22}^{(n)}\right); \tag{11}$$

$$b_{11}=\sum_{n=1}^{N}\tilde{\sigma}_{11}^{(n)}\Pi\left(\tilde{\sigma}_{11}^{(n)};\tilde{\sigma}_{22}^{(n)}\right); \qquad b_{22}=\sum_{n=1}^{N}\tilde{\sigma}_{22}^{(n)}\Pi\left(\tilde{\sigma}_{11}^{(n)};\tilde{\sigma}_{22}^{(n)}\right);$$

$$q=\sum_{n=1}^{N}\left(\left(\tilde{\sigma}_{11}^{(n)}\right)^2-\left(\tilde{\sigma}_{22}^{(n)}\right)^2\right);$$

$\tilde{\sigma}_{11}^{(n)}$, $\tilde{\sigma}_{22}^{(n)}$ – boundary values of the stress components σ_{11} and σ_{22}, which are found empirically.

3. Methods and algorithm for practical implementation of the mathematical model

According to the equation (6) and the system of equations (8), as well as the formulas (9) - (11) for determination and prediction of the boundary stress states of the orthotropic materials with the specified physical and mechanical characteristics, the following problems must be solved:

1) Solve the system of the equations (8);
2) Develop the algorithm for finding the roots of the nonlinear algebraic equation (6).

To simplify the procedure for solving the first problem, it should be noted, that since the third equation of the system (21) is linear, and the first two are independent of the parameter F, the problem of finding solutions of this system is equivalent to the problem of finding solutions of the system:

$$\begin{cases} \varphi_1(D,E)=0; \\ \varphi_2(D,E)=0, \end{cases} \tag{12}$$

The solution of the system of the equations (12) is based on a multi-step method [5], the algorithm of its practical implementation is to substantiate the initial approximate solution (D_0, E_0) of the problem and its further refinement by using the following iterative formulas:

$$Y^{(k)} = X^{(k)} - \frac{2}{3} J^{-1}(X^k) G(X^k); \tag{13}$$

$$X^{(k+1)} = X^{(k)} - \left(I - \frac{3}{2}\left(3J(Y^{(k)}) - J(X^{(k)})\right)^{-1} \right) \cdot \tag{14}$$

$$\cdot \left(3J(Y^{(k)}) - J^{-1}(X^k)\right) J^{-1}(X^k) F G(X^k)$$

where I – unity matrix; $X = (D,E)^T$ – vector of variables; $G(X) = (\varphi_1(X), \varphi_2(X))^T$ – vector-function; $J(X)$ – Jacobi matrix.

For determination of the approximate solution of the system (12), assume that the results of the experimental studies of orthotropic material on uniaxial tension and compression in the directions of anisotropy are known.

In particular, consider that under the uniaxial tension and compression conditions along one of the anisotropy directions $\sigma_{11} = \overset{+}{\sigma}_{11}$ and $\sigma_{11} = \overset{-}{\sigma}_{11}$, respectively, and along the other direction - $\sigma_{22} = \overset{+}{\sigma}_{22}$ and $\sigma_{22} = \overset{-}{\sigma}_{22}$.

Then, since according to the problem condition a pair of numbers $(\overset{+}{\sigma}_{11};0)$, $(0;\overset{+}{\sigma}_{22})$, and $(0;\overset{-}{\sigma}_{22})$ are approximate solutions of the equation (6), then D_0 and E_0 are approximate solutions of the following systems of equations:

$$\begin{cases} \dfrac{(1-\mu_{23}\mu_{32})}{2E_{11}} \overset{+}{\sigma}_{11} + \dfrac{D_0}{L} \overset{+}{\sigma}_{11} + \dfrac{F}{L} = 0; \\ \dfrac{(1-\mu_{23}\mu_{32})}{2E_{11}} \overset{-}{\sigma}_{11} + \dfrac{D_0}{L} \overset{-}{\sigma}_{11} + \dfrac{F}{L} = 0, \end{cases} \tag{15}$$

$$\begin{cases} \dfrac{(1-\mu_{13}\mu_{31})}{2E_{22}} \overset{+}{\sigma}_{22} + \dfrac{E_0}{L} \overset{+}{\sigma}_{22} + \dfrac{F}{L} = 0; \\ \dfrac{(1-\mu_{13}\mu_{31})}{2E_{22}} \overset{-}{\sigma}_{22} + \dfrac{E_0}{L} \overset{-}{\sigma}_{22} + \dfrac{F}{L} = 0, \end{cases} \tag{16}$$

From here:

$$D_0 = -\frac{(1-\mu_{23}\mu_{32})}{2E_{11}} L\left(\overset{+}{\sigma}_{11} + \overset{-}{\sigma}_{11}\right); \tag{17}$$

$$E_0 = -\frac{(1-\mu_{13}\mu_{31})}{2E_{22}} L(\sigma_{22}^{+} + \sigma_{22}^{-}).$$

So, the initial approximate solution of the system of equations (15) depends only on the physical and mechanical characteristics of the studied material.

For determination of the roots of the nonlinear equation (6) the method of continuous continuation by the best parameter is used [6, 7], its idea is to reduce the problem of finding solutions of the nonlinear algebraic equation of the second order (6) to the integration of the Cauchy problem:

$$\begin{cases} \frac{d\sigma_{11}}{d\lambda} = \mp \frac{f_{11}}{\sqrt{f_{11}^2 + f_{22}^2}}, \quad \sigma_{11}(0) = \sigma_{11}^{+}; \\ \frac{d\sigma_{22}}{d\lambda} = \pm \frac{f_{22}}{\sqrt{f_{11}^2 + f_{22}^2}}, \quad \sigma_{22}(0) = 0, \end{cases} \tag{18}$$

where

$$f_{11} = \frac{\partial\Phi(\sigma_{11},\sigma_{22})}{\partial\sigma_{11}}; \qquad f_{22} = \frac{\partial\Phi(\sigma_{11},\sigma_{22})}{\partial\sigma_{22}}; \tag{19}$$

$$d\lambda^2 = d\sigma_{11}^2 + d\sigma_{22}^2 \tag{20}$$

The independent variables σ_{11} and σ_{22} are functions of the parameter λ, which is calculated from the point $(\sigma_{11}^{+};0)$ along the tangent to the curve (6).

Thus, the algorithm for numerical implementation of the mathematical model (6) - (7) consists of:

1) Determining the model parameters D and E from the system of equations (12). For solving this problem, it is proposed to use a multi-step method, namely iterative formulas (13) - (14). The initial approximation of the solution of the system (12) is taken as numerical values D_0 and E_0 of the parameters D and E determined from the formulas (17);
2) Determining the numerical value of the model parameter F from the third equation of the system (8);
3) Drawing the curve (6) using the method of continuous continuation by the best parameter which consists in integration of the Cauchy problem (18) - (20) by the numerical Runge-Kutta method of the fourth order with the given integration step.

4. Results of mathematical modeling of boundary biaxial stressed states of wood and their analysis

In order to solve this problem, numerical experiments have been performed to determine the ultimate stresses in wood of different species.

In particular, in the MATLABR2012b environment the above proposed algorithm for studying the mathematical model (6) - (7) has been implemented and according to the Table 1 the delimitation curves of absolutely elastic and non-elastic deformation regions for pine tree with biaxial stress states in the planes of structural symmetry have been constructed.

Table 1. Physico-mechanical characteristics of wood at the temperature T = 20 °C and relative humidity W = 12% [8]: in the numerator - values of the tensile stresses, and in the denominator – values of the compression stresses

Wood species	Modules of elasticity, MPa			Boundaries of proportionality, MPa			Poisson's ratios		
	E_{aa}	E_{rr}	E_{tt}	σ_{aa}	σ_{rr}	σ_{tt}	$\frac{\mu_{ar}}{\mu_{at}}$	$\frac{\mu_{rt}}{\mu_{ra}}$	$\frac{\mu_{ta}}{\mu_{tr}}$
Pine	11300	1740	900	$\frac{102}{46.3}$	$\frac{10}{12}$	$\frac{7.8}{14}$	$\frac{0.078}{0.045}$	$\frac{0.306}{0.504}$	$\frac{0.463}{0.527}$

The results are graphically shown in Fig. 1.

IOP Conf. Series: Materials Science and Engineering 1016 (2021) 012001 doi:10.1088/1757-899X/1016/1/012001

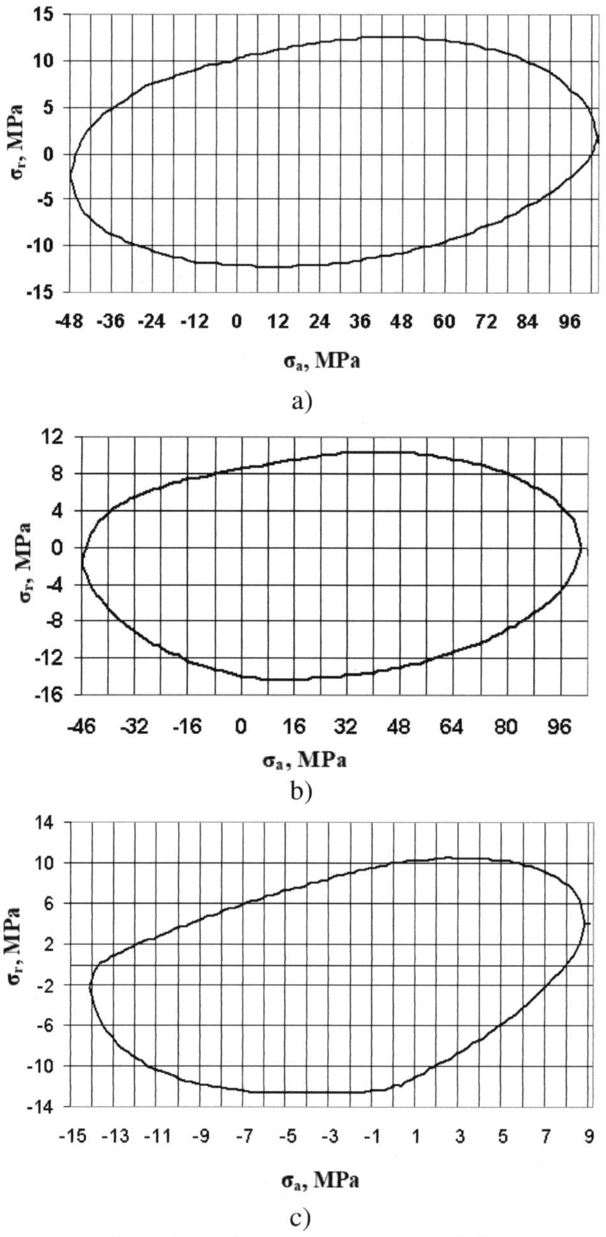

Figure 1. Delimitation curves of elastic and non-elastic areas of deformation of pine wood at the temperature T = 20 °C, relative humidity W = 12% and biaxial stress state in radial-axial (a), axial-tangential (b and radial-tangential (c) in planes of structural symmetry

The analysis of the obtained research results has been carried out and based on it, it is established that the characteristics of the drawn curves meet the basic heuristic requirements for the construction of known mechanical strength theories [1], according to them the curves of the ultimate stress state for anisotropic materials must:

1) cover the beginning of the Cartesian coordinate system;
2) be smooth and convex.

IOP Conf. Series: Materials Science and Engineering 1016 (2021) 012001 doi:10.1088/1757-899X/1016/1/012001

Conclusions

The new mathematical model for determining the elastic behavior region of anisotropic materials under the biaxial loadings has been developed, which, unlike known models, makes possible not only to describe satisfactorily, but also to substantiate the geometry of this region for materials, in which the asymmetry coefficients of the proportionality boundaries are different in different directions of anisotropy.

The adaptation of numerical methods, such as the method of continuous continuation by the best parameter, Runge-Kutta method and multi-step methods for construction of curves of delimitation of biaxial absolutely elastic and inelastic areas of pine deformation in planes of structural symmetry has been carried out.

The comparative analysis of the numerical and experimental data has been performed and it has been shown that they correlate well with each other. Thus, the adequacy of the proposed mathematical model and the correctness of the assumptions underlying this model have been confirmed.

References

[1] Bozhydarnyk V and Sulym H 1999 *Elements of the Theory of Plasticity and Strength* (Lviv: Svit **1** 532 in Ukrainian)

[2] Pobereyko S B Yakovenko A A Mysyk M and Kunynets Ye P 2017 The comparative analysis of theoretical studies results of the boundary stress state of anisotropic materials *Scientific Bulletin of UNFU* 27(9) 128-32

[3] Pobereyko S Holovatyy A Pobereyko B and Sokolovskyy Ya 2018 Mathematical model and software for determination of short-term strength of anisotropic composite materials *XIII International Scientific and Technical Conference on Computer science and informational technologies (CSIT)* 207-210

[4] Pobereyko S and Sokolovskyy Ya 2016 Mathematical model of anisotropic material strength under conditions of biaxial stress state *Scientific Journal of Lviv Polytechnic National University Series Computer Sciences and Information Technologies* **604** 148-53 (in Ukrainian)

[5] Zhmurko M and Krasnoshlyk N 2015 Comparison of multi-step iterative methods of higher orders for solving nonlinear equations and systems *Scientific Journal of Cherkasy University Series Applied Mathematics. Informatics* **18(351)** 1-8 (in Ukrainian)

[6] Mustafa O and Khirallah M 2013 Solving system of nonlinear equations using family of Jarratt methods *International Journal of Differential Equations and Applications* 12(2) 69-83

[7] Kuznetsov Ye and Leonov S 2017 Parametrization of the Cauchy problem for systems of ordinary differential equations with limiting singular points Journal of Computational Mathematics and Mathematical Physics **57(6)** 914-33 (in Russian) doi: 10.7868/S0044466918060066

[8] Borovikov A and Uholov B 1989 *Timber reference* (Moscow: Timber industry 296 in Russian)

IOP Conf. Series: Materials Science and Engineering 1016 (2021) 012002 doi:10.1088/1757-899X/1016/1/012002

Design of Parallel Sorting System Using Discrete-Time Neural Circuit Model

Pavlo V Tymoshchuk

CAD Systems Department, Lviv Polytechnic National University, Lviv, 79000, Ukraine

Email: pavlo.v.tymoshchuk@lpnu.ua

Abstract. A parallel sorting system based on discrete-time *K*-Winners-Take-All (KWTA) neural circuit (NC) model is designed. The system is described by a set of difference equations and by step functions. The KWTA NC model of discrete-time is given by difference equation and by step functions. Corresponding functional block-diagrams of the system are presented. The upper limit on the number of iterations required to achieve a convergence of the sorting process to the steady state is defined. The system does not need a knowledge of range of input data change. In order to use the system, the minimal difference between the inputs should be known. The system is suitable for processing unknown inputs of finite values located in arbitrary unknown finite range. The system is characterized by arbitrary finite resolution of inputs, high speed, moderate computational complexity and complexity of software implementation. The results of computer simulations illustrating efficiency of the system are provided.

1. Introduction

Sorting is an operation of arranging input data in an ascending ordered sequence, descending ordered sequence, alphabetic ordered sequence, etc. Over 25 percent of all computing time is spent on sorting. Sorting operation is widely applied in data and signal processing, telecommunication, navigation, mobile devices, artificial intelligence, very large scale integration design etc. [1]. Therefore, a problem of designing efficient sorting algorithms to minimize time and storage of sorting is extremely important.

A number of sequential sorting approaches are applied [1]. A sequential sorting algorithm may not be efficient sufficiently if a large volume of data should be sorted since they have quadratic computational complexity $\Theta(N^2)$. Therefore, parallel sorting algorithms of linear computational complexity $\Theta(N)$ can be used in this case. Various parallel sorting algorithms and architectures are discribed in [2]. Parallel sorting systems based on continuous-time neural networks (NNs) are applied in [3] – [10] and in references herein. Hardware implementations of sorting systems in VLSI technology are presented, for instance, in [8], [9].

Discrete-time neural networks are more accurate, reliable, more appropriate for processing real noised data. Such type of the networks are more convenient for implementation in parallel software. Moreover, they can be implemented in an up-to-date parallel digital hardware [11].

Let us design a parallel sorting system based on discrete-time neural circuit (NC) model presented in [12]. The system should possess arbitrary finite high resolution of input data, high speed of data processing, and linear computational complexity. Moreover, the system should have moderate hardware implementation complexity and be capable to process any finite unknown input data located in any finite unknown range. Computer simulations confirming theoretical derivations should be provided.

2. Basic discrete-time neural circuit

Consider a vector $a=(a_1,a_2,...,a_N)^{\mathrm{T}} \in \Re^{\mathrm{n}}$, $1<N<\infty$ of real input data with unknown values of its elements. The inputs are distinct and finite, i.e.

$$a_i \neq a_j,\ -\infty < a_i, a_j < \infty, \tag{1}$$

where $i \neq j = 1,2,...,N$. The simple and fast discrete-time K-winners-take-all (KWTA) NC presented in [12] is capable to determine the largest K of these inputs, which are appealed to as the winners. In particular, the circuit processes the vector $\boldsymbol{a}$ of these inputs to get a such vector $b=(b_1,b_2,...,b_N)^{\mathrm{T}}$ of corresponding outputs that the following KWTA property is met:

$$b_i > 0, i = 1,2,...,K;\ b_j < 0, j = K+1, K+2,...,N. \tag{2}$$

We simplify the NC presented in [12] having described it by the following difference equation:

$$y(l+1) = y(l) + r\,sgn(D(y)), \tag{3}$$

where $y(l)$ is a discrete-time state variable with an initial condition $-\infty < y(1) < \infty$, $D(y) = \sum_{k=1}^{N} S_k(y) - K$ is a difference function between obtained and required numbers of positive outputs,

$$S_k(y) = \begin{cases} 1, & if \quad a_k - y > 0; \\ 0, & otherwise \end{cases} \tag{4}$$

is a step function,

$$sgn(D(y)) = \begin{cases} 1, & if \quad D(y) > 0; \\ 0, & if \quad D(y) = 0; \\ -1, & if \quad D(y) < 0 \end{cases} \tag{5}$$

is a signum function, r is a parameter that should have a value less than minimal difference between inputs, i. e., the inequalities $0 < r \leq min|a_i - a_j|$, $i \neq j = 1,2,...,N$ should be met. The quantity of iterations required to reach a convergence of the search process to the steady state KWTA operation is limited from above as follows:

$$m \leq |y(1) - y^*| / r, \tag{6}$$

where y^* is a steady state value of $y(l)$. As it can be seen, the NC described by difference equation (3) and by output equations (4) requires setting a value of the parameter r that should be smaller than minimal difference between inputs. This means that the minimal difference between inputs should be known.

3. A discrete-time model of neural circuit based parallel sorting system

In parallel sorting, an order of sorting can be represented as a permutation matrix. In such a matrix, "1" in the row marked with a_i and column labeled with g_j can be determined as the ith item in an unsorted list and jth item in a sorted list [8] – [10]. For instance, in the case $i=1,2,...,10$ an unordered list of inputs $\{a_1,a_2,...,a_{10}\}$ and their sorted list $\{a_2, a_4, a_1, a_3, a_5, a_7, a_6, a_{10}, a_8, a_9\}$ are described by the following corresponding permutation matrix:

	d_1	d_2	d_3	d_4	d_5	d_6	d_7	d_8	d_9	d_{10}	*rank*
a_1	0	0	1	0	0	0	0	0	0	0	3
a_2	1	0	0	0	0	0	0	0	0	0	1
a_3	0	0	0	1	0	0	0	0	0	0	4
a_4	0	1	0	0	0	0	0	0	0	0	2
a_5	0	0	0	0	1	0	0	0	0	0	5
a_6	0	0	0	0	0	0	1	0	0	0	7
a_7	0	0	0	0	0	1	0	0	0	0	6
a_8	0	0	0	0	0	0	0	0	1	0	9
a_9	0	0	0	0	0	0	0	0	0	1	10
a_{10}	0	0	0	0	0	0	1	0	0	0	8.

(7)

The matrix (7) can be converted to the sorting matrix given by

	S^1	S^2	S^3	S^4	S^5	S^6	S^7	S^8	S^9	S^{10}	*rank*
a_1	0	0	1	1	1	1	1	1	1	1	3
a_2	1	1	1	1	1	1	1	1	1	1	1
a_3	0	0	0	1	1	1	1	1	1	1	4
a_4	0	1	1	1	1	1	1	1	1	1	2
a_5	0	0	0	0	1	1	1	1	1	1	5
a_6	0	0	0	0	0	0	1	1	1	1	7
a_7	0	0	0	0	0	1	1	1	1	1	6
a_8	0	0	0	0	0	0	0	0	1	1	9
a_9	0	0	0	0	0	0	0	0	0	1	10
a_{10}	0	0	0	0	0	0	0	1	1	1	8

(8)

In general case the sorting results can be presented as follows:

$$d_1 = a^{\mathrm{T}} S^1,\ d_{K+1} = a^{\mathrm{T}} (S^{K+1} - S^K), \tag{9}$$

where the elements of the Kth column $S^K = \left[S_1^K, S_2^K, ..., S_N^K\right]^{\mathrm{T}}$, $K=1,2,\ .\ .\ .\ ,\ N-1$ of the sorting matrix are determined by the step functions (4) using the difference equation (3).

Since $S^N = [1,1,...,1]^{\mathrm{T}}, S^N = [1,1,...,1]^T$, only $N-1$ equations (3) are necessary and each of these equations computes one column of the sorting matrix from left to right with k increasing from 1 to $N-1$. Therefore, only $N-1$ neurons are required when compared with the other sorting systems that use N^2 neurons [9]. Specifically, the 1WTA network is necessary to define the largest element of the list. The 1WTA network and the 2WTA network are required for computing the second item in the list in a parallel mode without recalculation the first item. The 2WTA network and the 3WTA network are necessary for determining the third item in the list in a parallel mode without recalculating the second item and so on. As such, the whole list of N items can be sorted by using the KWTA, $K=1,2,...,N-1$ networks without the need to calculate the last item by setting $S^N = [1,1,...,1]^{\mathrm{T}}$ [3], [8].

4. Functional block-diagram of the system

A functional block-diagram of the discrete-time NC described by difference equation (3) and by output equations (4) is shown in figure 1(a). Figure 1(b) presents a functional block-diagram of part of parallel sorting system described by thebfirst equation of (9). A functional block-diagram of part of parallel sorting system described by the second equation of (9) is shown in figure 1(c). The diagrams include inputs $a_1, a_2, ..., a_N$, summers $\sum$, blocks $S_1, S_2, ..., S_N$ of step functions $S_k(y)$, $k=1,2,3,...,N$, signum function sgn (5), multiplier $\times$, time delay z^{-1}, external sources of constant signals $r, K, y(1)$, and outputs $S_1(y), S_2(y), ..., S_N(y)$.

IOP Conf. Series: Materials Science and Engineering 1016 (2021) 012002 doi:10.1088/1757-899X/1016/1/012002

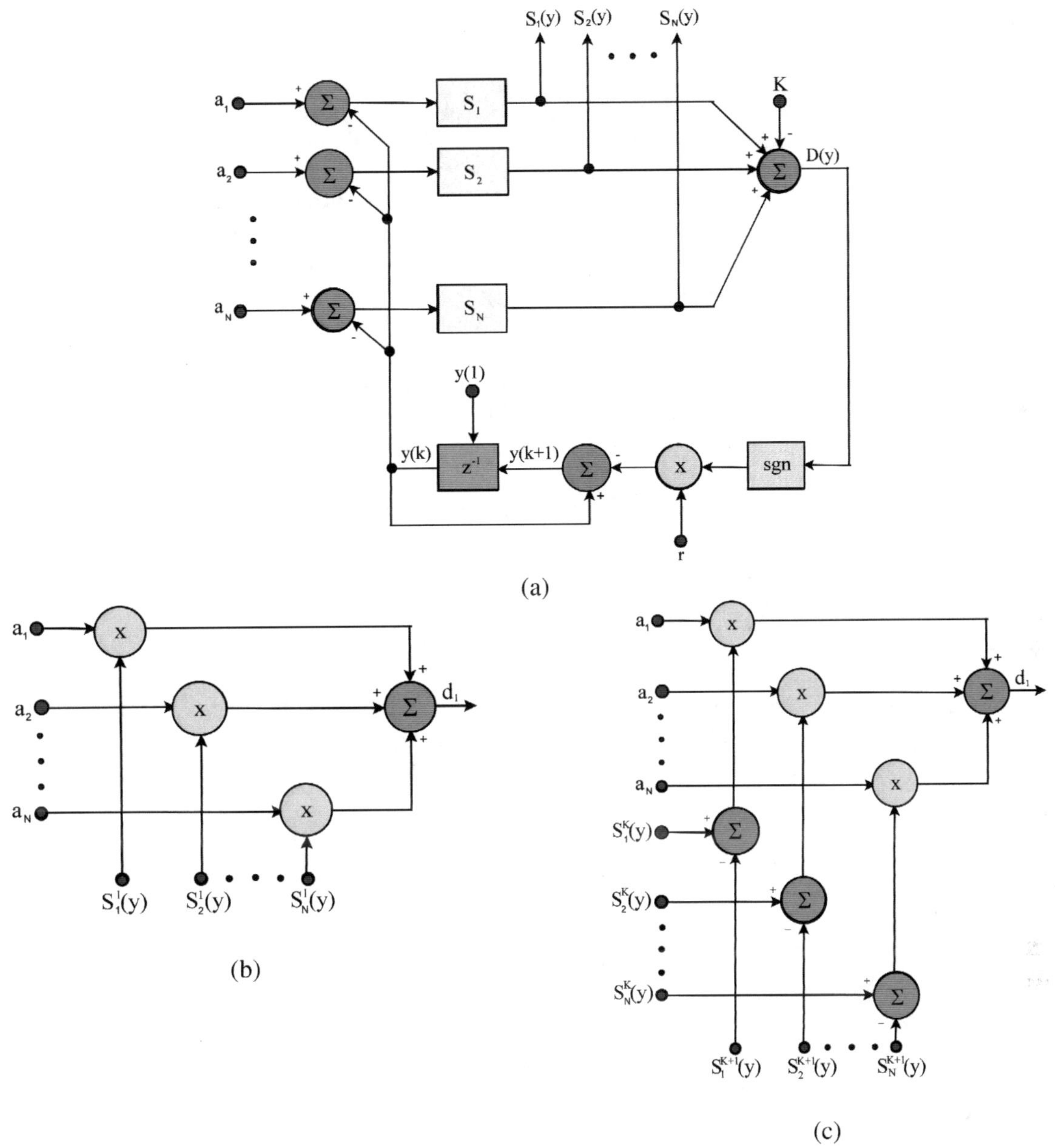

Figure 1. (a) The functional block-diagram of the discrete-time NC described by difference equation (3) and by output equations (4); (b) The functional block-diagram of part of parallel sorting system based on NN described by first equation of (9); (c) The functional block-diagram of part of parallel sorting system based on NN described by second equation of (9).

The diagram shown in figure 1(a) consists of one multiplier, one time delay unit, three sources of constant signals (or two sources of constant signals if $x(1)=0$), $N+1$ controlled switches, and $N+2$ summers. The diagram presented in figure 1(b) has N multipliers and one summer. The diagram shown in figure 1(c) consists of N multipliers and $N+1$ summers.

A functional block-diagram of one of the simplest comparative discrete-time parallel sorting system based on continuous-time KWTA NNs with the Heaviside step activation function and a single state variable described in [10] that was applied for parallel sorting in similar way needs one source of constant signals, one integrator, N switches, and $N+1$ summers. Thus, from the complexity point of view, the discrete-time parallel sorting system based on NC described by difference equation (3) and by output equations (4) is close to other comparable analogs.

The functional block-diagrams presented in figure 1 can be implemented in a parallel software or parallel digital VLSI hardware. Software implementation of the diagrams can be preferred if it is necessary to have an accurate, simple and flexible parallel realization. The diagrams can be implemented in a digital hardware if fast, compact, reliable, and noise insensitive realization is required [13].

5. Computer simulations of the system

Consider an example of computer simulations that demonstrate the performance of the discrete-time parallel sorting system described by equations (3), (4), and (9). Let us sort nine elements of uniformly distributed on the interval *(-10.0, 10.0)* random numbers, using parallel sorting system described by equations (3), (4), and (9). We set the initial values of the discrete-time state variables $x^k(1)=5.0$, $k=1,2,...,N$. Using eight NCs described by difference equations (3) and by output equations (4) we need only eight neurons unlike to the *81* neurons in the analog sorting network in [9].

Figure 2 shows the dynamics of inputs and outputs of the system used for parallel sorting with $r=0.01$, $y^k(1)=5.0$, $k=1,2,...,N$. As it can be seen, the correct sorting of the inputs is reached in $n<40<\frac{|0-5|}{0.01}=500$ iterations in accordance with estimate (6). Elapsed time necessary for completing this sorting in Matlab is approximately equal to *0.07 s*.

Figure 3 presents the maximal numbers of iterations required for parallel sorting nine elements of uniformly distributed on the interval *(-10.0, 10.0)* random numbers using parallel sorting system described by equations (3), (4), and (9) for $r=0.01$ and for different initial values of the discrete-time state variables $x^k(1)$, $k=1,2,...,N$ chosen on the interval *(-10.0, 10.0)*. As it can be seen in figure 3, the maximal number of iterations necessary to sort the inputs in this case is not larger than 50 for all values of $x^k(1)$, $k=1,2,...,N$. Maximal elapsed time required for completing this sorting in Matlab is approximately equal to *0.05 s*.

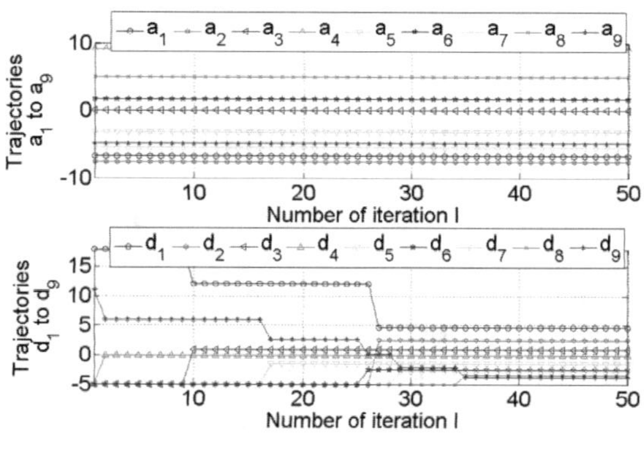

Figure 2. Trajectories of outputs of the parallel sorting system described by equations (3), (4), and (9) for *N=9* uniformly distributed on the interval *(-10.0,10.0)* random inputs *a*, $r=0.01$, and $y^k(1)=5.0$, $k=1,2,...,N$.

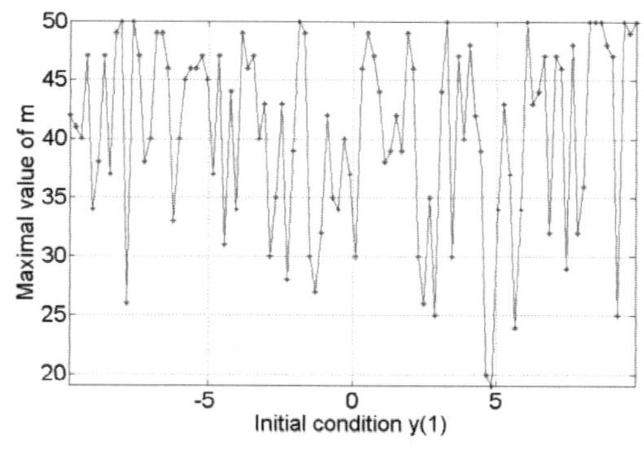

Figure 3: Maximal numbers of iterations necessary for parallel sorting by the parallel sorting system described by equations (3), (4), and (9) for $r=0.01$ and for different initial values of the discrete-time state variables $x^k(1)$, $k=1,2,...,N$ chosen on the interval *(-10.0,10.0)*.

As it can be seen from the results of simulations, the parallel sorting system described by equations (3), (4), and (9) is accurate and fast. Moreover, this system does not need a knowledge of change range of inputs. In order to use the system a minimal difference between inputs should be known like close comparative discrete-time analog presented in [10]. However, in contrast to parallel sorting system described by equations (3), (4), and (9), this analog requires multiplications of large value difference functions by small value parameters in a right-hand side of the difference equation describing discrete-time KWTA NN model in the case of large number of inputs and small minimal difference between them. This can cause arising inaccuracies which lead to incorrect operation of such parallel sorting system. In addition, indicated multiplications can be difficult to implement in a corresponding hardware.

Conclusions

The discrete-time parallel sorting system of arbitrary unknown finite inputs located in unknown finite change range and corresponding functional block-diagrams are presented. The system possesses high operation speed and arbitrary finite resolution of inputs. Practical operation speed of the system and its resolution are limited by restrictions of software or hardware implementation of the system. The system can be implemented in a parallel software or parallel digital VLSI hardware. The software implementation can be chosen if an accurate, simple and flexible parallel realization is required. The system can be implemented in a digital hardware for parallel operation if fast, reliable, and noise insensitive realization is necessary. High performance of the system used for fast and accurate parallel sorting has been demonstrated by computer simulations. The system can be recommended to use if a large number of unknown finite inputs with known minimal difference between them located in unknown finite change range should be sorted with high resolution and speed.

References

[1] Knuth D E 1973 *The Art of Computer Programming, Sorting and Searching* (Reading: Addison-Wesley)
[2] Akl S G 1985 *Parallel Sorting Algorithms* (Orlando: Academic)
[3] Atkins M 1989 *Proceedings of International Joint Conference on Neural Networks (Washington DC)* (Washington DC: IEEE) p 65
[4] Takefuji Y and Lee K S 1990 *IEEE Transactions on Circuits and Systems* **CAS-37** **11** 1425-29
[5] Chen W and Hsieh K 1990 *Proceedings of International Joint Conference on Neural Networks (San Diego)* vol. III (San Diego: IEEE) p 793
[6] Kwon T M and Zervakis M 1992 *Proceedings of International Joint Conference on Neural Networks (Baltimore)* vol. I (Baltimore: IEEE) p 701
[7] Tseng Y H and Wu J L 1992 *Electronics Letters* **28** **10** 906-08
[8] Wang J 1995 *IEEE Transactions on Neural Networks* **6** **4** 962–71
[9] Kwon T M and Zervakis M 1995 *Multidimensional Systems and Signal Processing* **6** **4** 333-46
[10] Wang J 2010 *IEEE Transactions on Neural Networks* **21** **9** 1496-06
[11] Cichocki A and Unbehauen R 1993 *Neural Networks for Optimization and Signal Processing* (New York: Wiley)
[12] Tymoshchuk P V 2009 *Neurocomputing* 72 3191-02
[13] Tymoshchuk P V and Shatnyi S V 2015 *Proceedings of XX International Seminar/Workshop DIPED* (L'viv: IAPMM Press) p 168

IOP Conf. Series: Materials Science and Engineering 1016 (2021) 012003 doi:10.1088/1757-899X/1016/1/012003

PCB Open, Short and Shift Defects Detection by Logical Comparison of Mask Images

R A Melnyk* Yu V Havrylko and Ya V Repa

Lviv Polytechnic National University, Lviv, Ukraine

*Email: ramelnyk@polynet.lviv.ua

Abstract. The known skeletonization algorithm was realized and the skeleton specific points were found. The algorithm was applied to printed circuit boards to detect the main defects: wire and contact breaks, short circuits. Mask images on a base of specific points and logical comparison to select and to mark places with defects were realized. The example of applications is shown.

1. Introduction

Mechatronics, robotics and many other automation systems could not be projected without usage of electronic control devices. The last ones are based mainly on PCB and chips. All manufacturing processes have probability of making defects in the product. Algorithms of computer vision in this case help to detect and to eliminate defects increasing reliability and duration of a mechanical system work. An open connection is the main defect on the PCB and is caused by a lack of contacts within the lead, between the lead and pad. Shorts occur when solder is connecting one lead to another. Component shift is being described as the misalignment of constructive elements to their designed positions. These three defects occupy two third of all defects. In the paper [1] the proposed algorithm combines parallel and sequential processes in iterative procedure. The proposed method conducted into experiments of many benchmark dataset were used to evaluate the algorithm robustness. The article [2] a quantum version of Hilditch's algorithm is proposed where the quantum images were used. The paper [3] considers five stages of the algorithm: Image rebuilding, inspection image normalization, image subtraction, defects separation and defect classification. The s system recognizes different types of defects, for example, open, pinhole, missing conductor and so on. The given method produces good results. In the article [4] classification of the defects on PCBs was made by a hybrid algorithm: a morphological image segmentation algorithm and simple the image processing. Also, good results were obtained. In the paper [5] topological information on PCBs is introduced. The topological comparison method compares the inspection graph obtained from the skeletons of the conductor and insulator images. The approach in the article [6] is based on the comparison between the standard PCB image and the target image. Multiple images of the qualified PCBs are used at the same position. An average of the images is taken as an initial standard image. The article presents the algorithm of the etalon and real PCB image skeletons comparison by logical operation with mask images taking into account a tolerance for a distance between suspicious places.

2. Image pre-processing

An example of the PCB image from [10] is taken for experiments. There are three colors in Fig.1: dark green background, light green routes and light gold contacts. This image is being transformed to a binary form because preliminary conditions of the main algorithm need to input the PCB image with a white background and black routes and contacts.

IOP Conf. Series: Materials Science and Engineering 1016 (2021) 012003 doi:10.1088/1757-899X/1016/1/012003

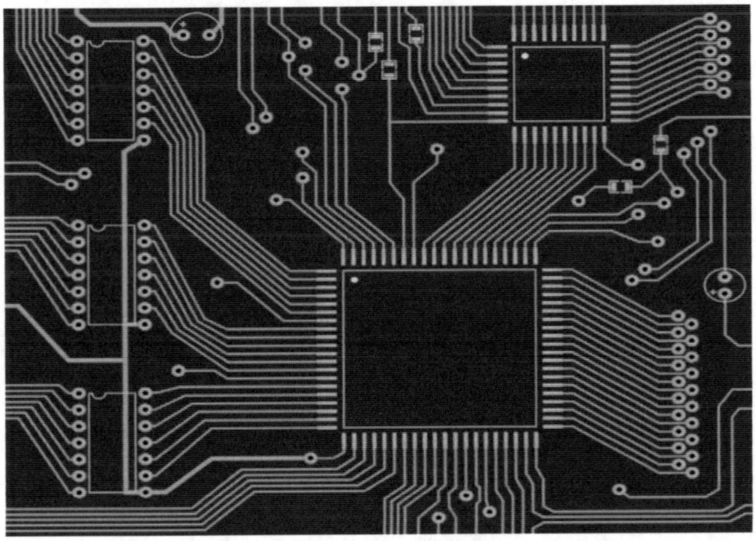

Figure 1. Image of PCB

For pixels of the PCB image a cumulative histogram is being calculated:

$$G(s) = \{\sum g(j), j = 0, \ldots, s\}, s = 0, \ldots, 255 \tag{1}$$

where $G(s)_i$ is a value of the cumulative frequency in the s-th interval, g(j) is a number of pixels in the j-th interval of intensity.

Images of manufactured PCBs differ between themselves by the reason of irregular lightning, changing background etc. Thus, the colors of the background of the PCB and routing wires and contacts are not uniform.

This fact is being confirmed by a cumulative histogram which plot in Fig.2 ununiformly increases reflecting changes of pixel intensity

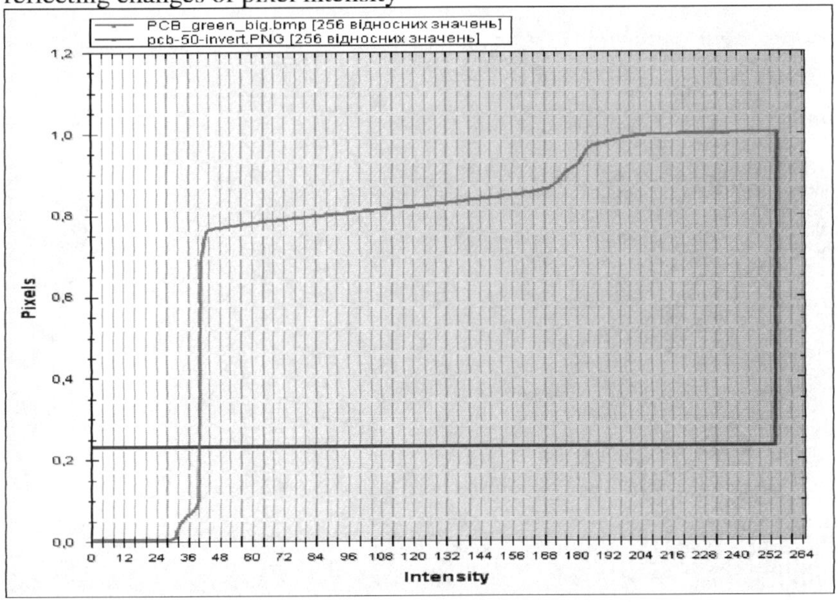

Figure 2. Cumulative histograms of original and binarized of the PCB images

A cumulative histogram is being approximated by piecewise-linear function $g(x)$ determined on the interval $a_i \le x_i \le a_{i+1}$ with a set of sections having the end points (α_i, β_i) and $(\alpha_{i+1}, \beta_{i+1})$. The function $g(x)$ minimizes the approximation error ε got by the following formula:

$$\varepsilon = \max_{a_i \le x \le a_{i+1}} |g(x) - f(x)| \tag{2}$$

For piecewise-linear approximation of the histogram function the realized Ramer–Douglas–Peucker (RDP) algorithm from [11] was taken as basic.

To find a threshold value for the PCB image in Fig.1a method of multilevel segmentation based on linear piecewise linear approximation of a cumulative histogram is being used. A result of approximation is shown in Fig.3. The main threshold value is 45 which we increase by 5 points to 50 and transform the image to binary form replacing pixels in the interval 0-50 by black color and in the interval 51-255 by white pixels. Then for the resulting image of PCB cumulative histogram is given in Fig. 1 (flat chart). To satisfy the input conditions of the skeletonization algorithm the resulting binary image is being inverted (Fig.4).

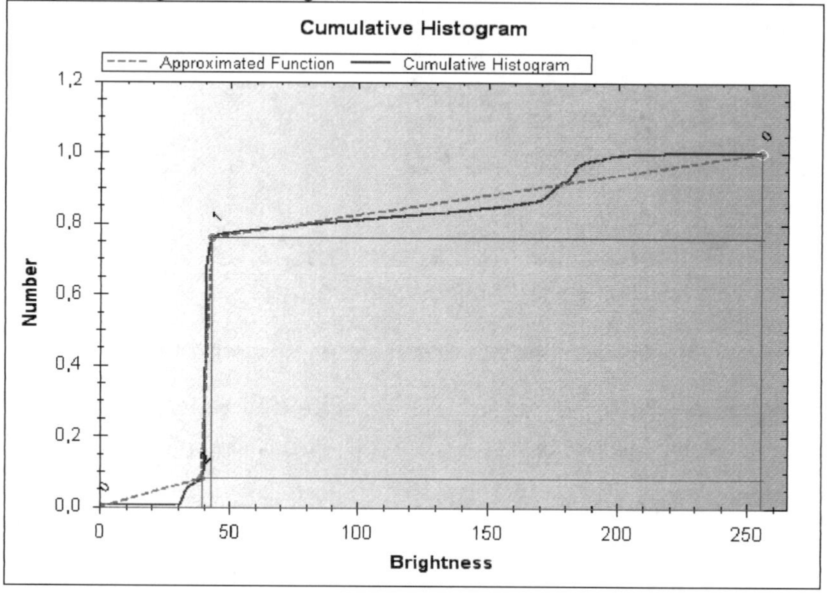

Figure 3. Cumulative histogram and its piecewise approximation

The resulting image from Fig.4 is being taken as input data for the skeletonization algorithm which gives the skeleton image in Fig.5.

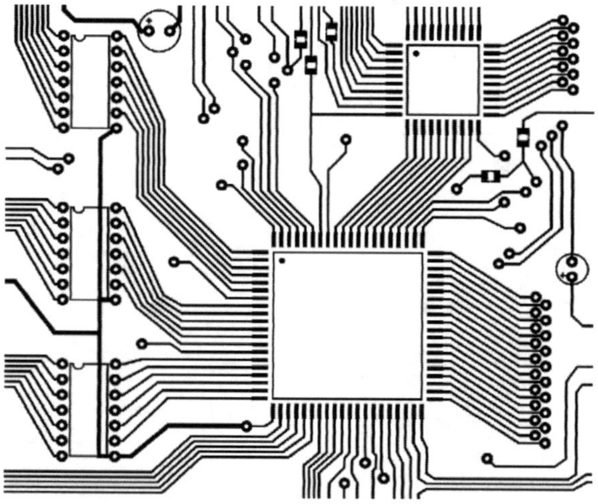

Figure 4. Binarized PCB image.

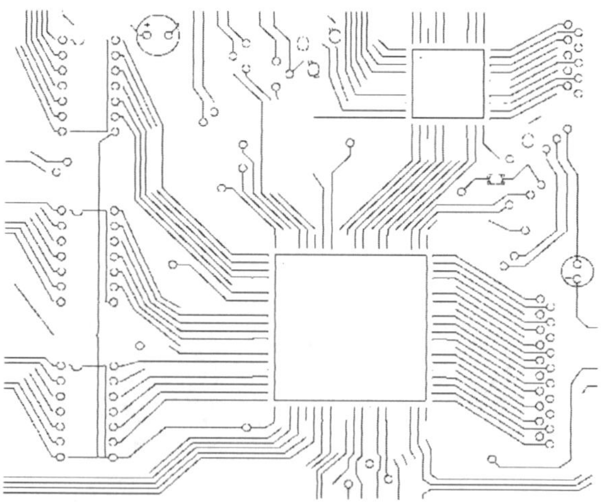

Figure 5. Skeleton of PCB image.

Having a technique for an image evaluation, segmentation, inversion and transformation to skeletons it is possible to widen these services to obtain more practical instruments.

3. PCB skeleton and its specific points

The skeleton of a binary image [7,8,9] was used to analyse the components of printed circuit boards to find irregularity and defects of routing wires and their groups of contacts. In Fig 5 an example of the PCB image skeleton is given. For it the endings and switches are found and marked by blue and red rectangles as it is shown in Fig.6. A switch indicates connections of wires with contacts or between themselves. On the skeleton endings indicate coordinates of rectangular contacts and wire breaks.

There are two skeletons in Fig.6: one has marked switches by blue rectangles and the second red rectangles indicate endings. There are no defects on these skeletons. Switches are native for the original PCB image and endings on the second skeleton are also native and mark breaks of routes on the edge of the image and the ends of the contacts. In common, endings on the edge could not be considered.

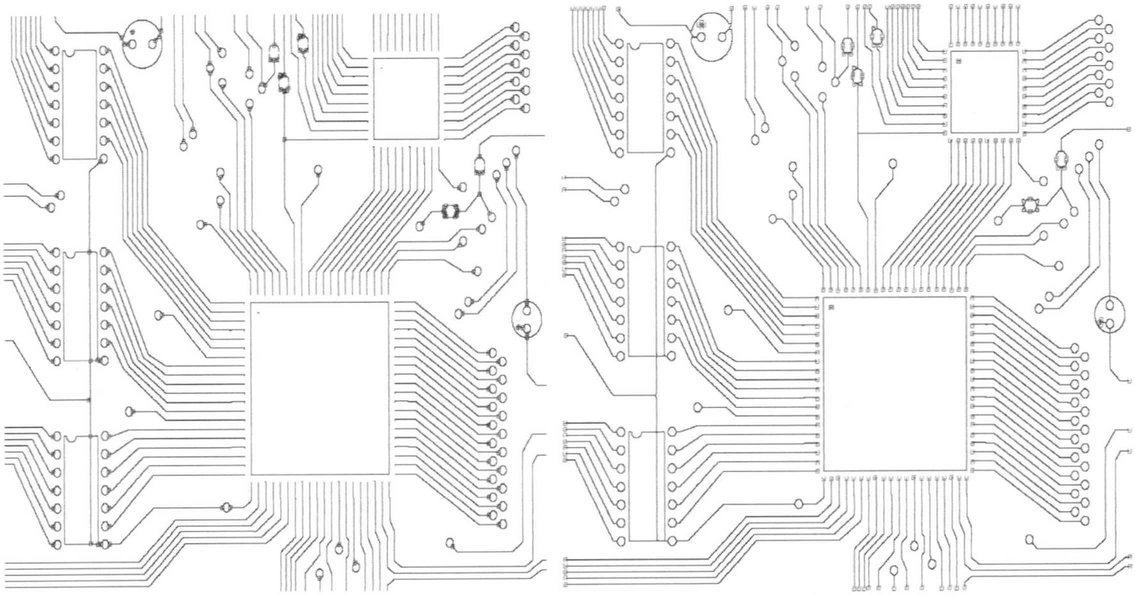

a b

Figure 6. Etalon PCB skeletons with specific points of switches (a) and endings (b)

4. PCB open and short detection

In Fig.6 specific points of endings and switches do not always indicate defects. They characterize some constructive elements needed for normal work of the circuit. Having the skeleton and coordinates of its specific points it is impossible to find defects by only one real sample of the PCB image. For specific points classification as defects or robust elements of the PCB it is needed to build the image skeleton, to find its specific points, to repeat these manipulations with the etalon PCB image and to take decision about types and placement of defects.

After building two skeletons for the PCB etalon and real PCB images there are four sets of specific points: for the skeleton of the etalon (reference) PCB image a set of endings $E_r=\{e_{r1}, e_{r2},\ldots,e_{rn}\}$ and a set of switches $W_r=\{w_{r1}, w_{r2},\ldots,w_{rm}\}$ and for the skeleton of the real PCB image a set of endings $E_d=\{e_{d1}, e_{d2},\ldots,e_{dn}\}$ and a set of switches $W_d=\{w_{d1}, w_{d2},\ldots,w_{dm}\}$.

Elements of these sets are coordinates of different detected specific points, for example, $e_{r1}=(x_{r1}, y_{r1})$, $w_{d1}=(x_{d1}, y_{d1})$ etc. There are two fragments of skeletons in Fig.7 obtained from the PCB image having two breaks of wires. First of them has only switches and one specific switch near contact is missing due to the break of wire near contact. This case is illustrated by Fig.7a the second fragment in Fig.7b demonstrates three endings for two breaks of wires.

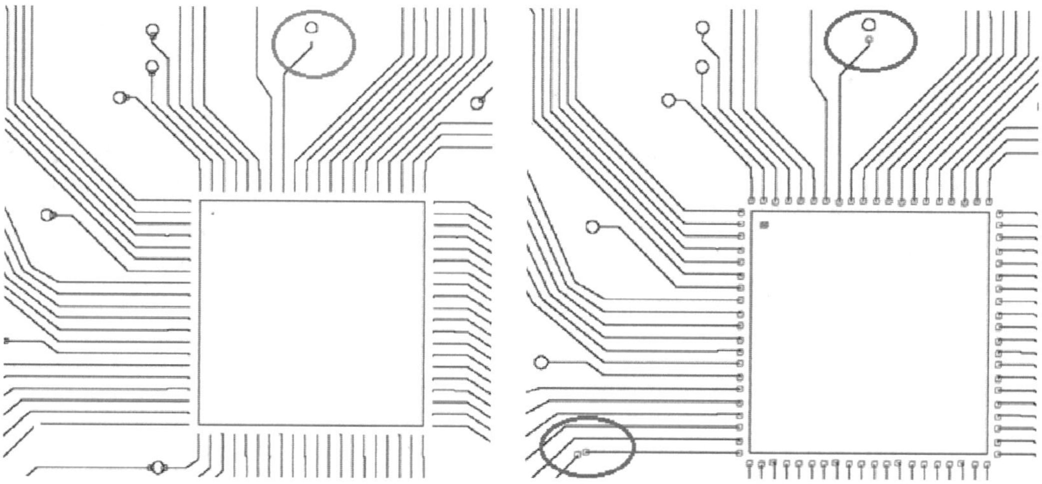

Figure 7. Fragments of skeleton images with switches (a) and endings (b) of the PCB having two breaks

If the PCB image has many elements, wires and contacts it is difficult by a visual observation to separate rectangles related to defects from those belonging to correct elements. An idea how to detect, to mark and to select places with defects on the PCB area is considered. The procedure is based on four sets of specific points given by their coordinates. Four images are being built: two for the etalon PCB and two for the real manufactured PCB. They are called masks: M_{ee}, M_{es}, M_{re}, M_{rs}. Every mask image has dimension of the corresponding PCB image. Black circles (in our case for switches blue and for endings red) are being drawn on a white or transparent background. Centers of circles have coordinates of found specific points: endings or switches. A radius of a circle reflects a manufacturing precision. Its value is being taken from the interval of 0 - 5 pixels. Two examples of the etalon mask and the real PCB mask are shown in Fig.8. points.

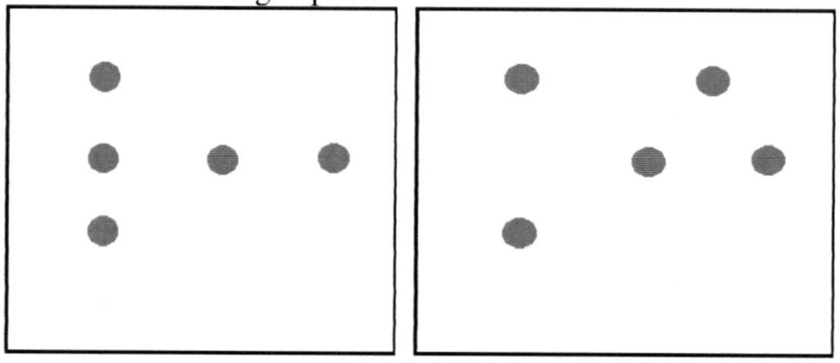

Figure 8. Etalon mask (a) and mask for PCB with defects (b)

For every pair of masks M_{ee}, M_{es} and M_{re}, M_{rs} logical operations are being realized:

$$\begin{cases} M = M_{ee}\ XOR\ M_{es} \\ M = M_{re}\ XOR\ M_{es} \end{cases} \quad (3)$$

At the beginning of these operations the etalon mask is being accepted as the resulting image. In these operations a pixel of the resulting image changes its color if two components of the operation have different colors. That is if a pixel was black it becomes white and if it was white it becomes black. If pixels of two components have the same color the resulting pixel remains as white or accepts white color being before as black. The resulting image for operations (3) upon images from Fig.8 is shown in Fig.9a.

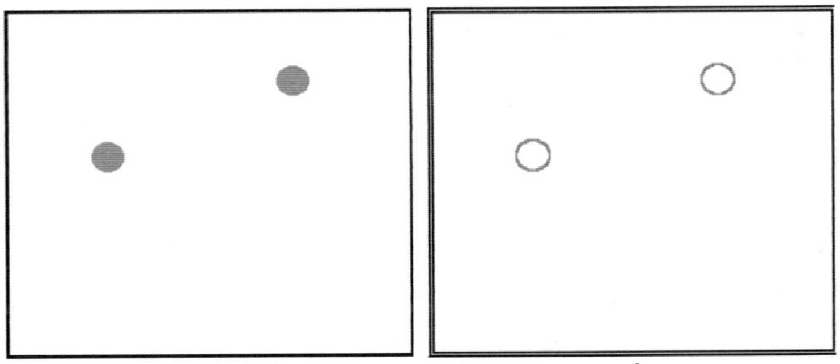

Figure 9. Resulting mask (a) and its edges (b)

In practice centers of circles on the etalon PCB and the real PCB masks do not coincide exactly and the resulting image contains figures having a form of intersected circles two examples of which are shown in Fig.10. These examples illustrate a case of two switches: one of the etalon masks and one of a real image mask A size of the intersection area is inversely proportional to a distance between centers of corresponding circles on the etalon and real masks.

Figure 10. Two intersected circles with different distances between them

Intersected parts are marked by white color or are being assigned to a transparent property. Non intersected parts have two possibilities: they accept colors of initial circles or become transparent. A criterion for elimination of non-intersected parts of circles is being formed by a relation between an area of the intersected part and an area of the full circle:

$$S_i/S_f < Tol \quad (4)$$

where S_i, S_f are squares of an intersected part and a full circle, Tol is a relative tolerance value to control a distance between corresponding elements of the etalon PCB and the controlled samples. Practically this parameter reflects a number of pixels by which real PCB image is shifted from the etalon PCB image.

After performing logical operations of the formulas (1) for switches or ending we obtain four types of mask objects: 1) circles when the etalon mask have no a switch and the real mask has a switch (or vice versa), 2) intersected circles when the etalon mask and the real mask have switches shifted between themselves, 3) intersected circles with non-marked intersected part when two intersected circles are only on the etalon mask or only on the real mask and, 4) circles without intersected part when to the case in point 3 opposite mask has one circle. Examples of these types are shown in Fig.11

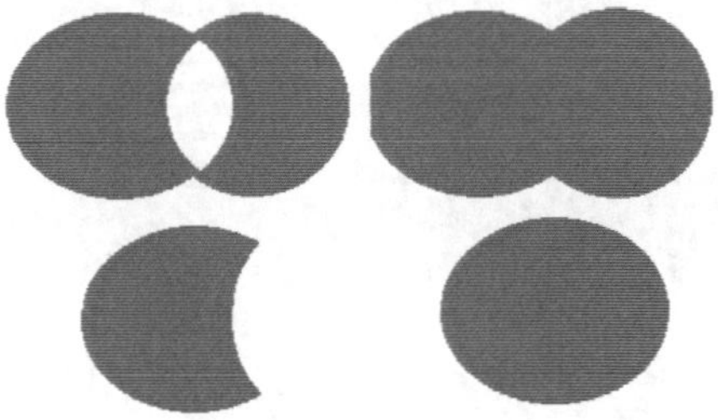

Figure 11. Types of figures after logical comparison of masks

Presented figures happens with the highest frequency. Theoretically intersected figures could have more complicated form because places of defects and shifting of their position are unpredictable.
To make obtained figures transparent for defects observation an algorithm for edge detection was applied. The resulting images are of two colors and a filter based on the following matrix for convolution operations is used:

$$EM = M\ x\ M_c$$

where

$$M_c = \begin{pmatrix} -1 & -1 & -1 \\ -1 & 8 & -1 \\ -1 & -1 & -1 \end{pmatrix} \tag{5}$$

The borders in Fig.12 of indicate places containing suspicious defects caused by open, short or shift errors.

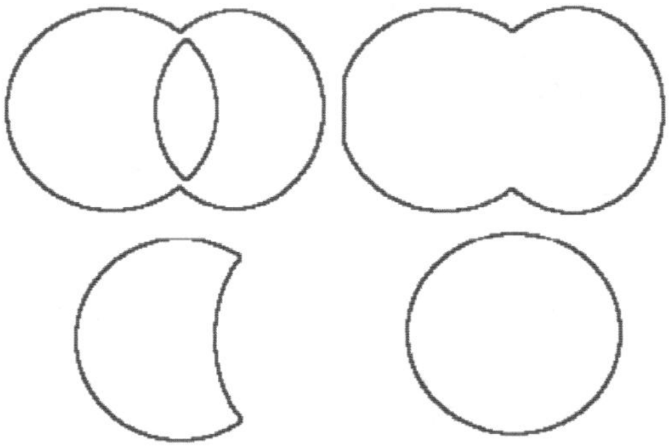

Figure 12. Types of edges

The tolerance value also has a function to control the drawing process: to show these borders on the resulting image or to hide them. The edge mask image EM is being imposed on the original manufactured PCB image to underline found and suspicious defects that is demonstrates on the example in Fig.13.

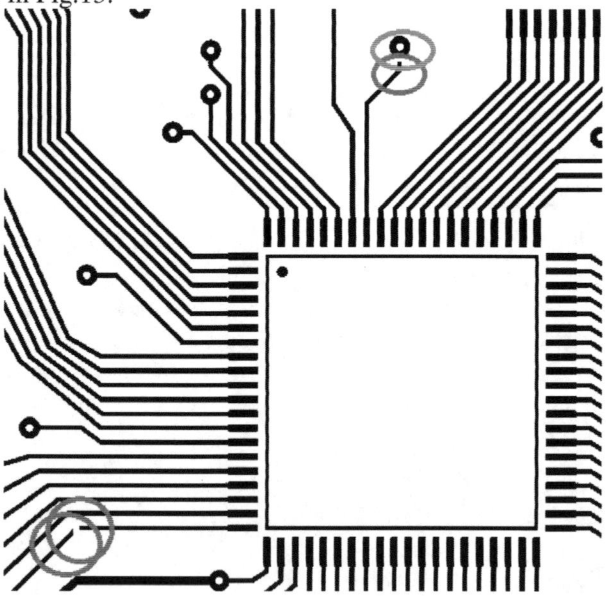

Figure 13. PCB with marked defects of switches and endings

So, researches have an opportunity to investigate PCB images by two control parameters: diameters of circles and the tolerance value of shifting.

Resulting images of five stages given by developed software are demonstrated in Fig.14.

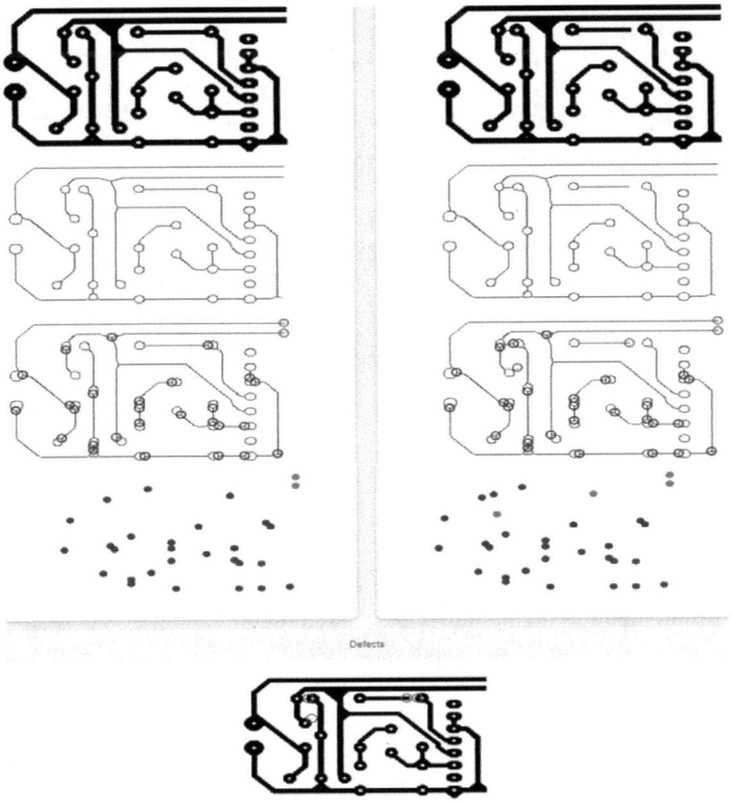

Figure 14. Images of the different steps of the algorithm

Conclusions

The Hilditch's skeletonization algorithm was realized and used for binary PCB images to find two types of feature points: endings and switches of skeleton. First ones indicate the wire breaks and the second ones indicate short circuits. Mask images with suspicious places of defects were formed and logical XOR operations to manipulate them were used. The developed software can be used for PCB defects detection in automated system of control devices design and manufacturing.

References

[1] Abu-Ain W Abdullah S N H S Bataineh B Abu-Ain T and Omar K 2013 Skeletonization algorithm for binary images *Procedia Technology* **11** 704-709

[2] Naseri M Heidari S Gheibi R Gong L H Rajii M.A and Sadri A 2017 A novel quantum binary images thinning algorithm: A quantum version of the Hilditch's algorithm *Optik* **131** 678-86

[3] Morrison P and Zou J J 2005 An effective skeletonization method based on adaptive selection of contour points *Third International Conference on Information Technology and Applications (ICITA)* **1** 644-49

[4] Rau H and Wu C H 2005 Automatic optical inspection for detecting defects on printed circuit board inner layers *The International Journal of Advanced Manufacturing Technology* **25(9-10)** 940-6

[5] Putera S I and Ibrahim Z 2010 Printed circuit board defect detection using mathematical morphology and MATLAB image processing tools *2nd International Conference on Education Technology and Computer* **5** V5-359

[6] Ito M and Nikaido Y 1991 Recognition of pattern defects of printed circuit board using topological information *Eleventh IEEE/CHMT International Electronics Manufacturing Technology Symposium* 202-206

[7] Hilditch's Skeletonization Algorithm
http://cgm.cs.mcgill.ca/~godfried/teaching/projects97/azar/skeleton.html

[8] Zhang T Y and Suen C Y 1984 A fast parallel algorithm for thinning digital patterns *Communications of the ACM* **27(3)** 236-9

[9] Lee T C Kashyap R L and Chu C N 1994 Building skeleton models via 3-D medial surface axis thinning algorithms *CVGIP: Graphical Models and Image Processing* **56** 462-78

[10] Why Do We Use Copper to Make PCB Traces?
https://yic-assm.com/why-do-we-use-copper-to-make-pcb-traces/

[11] Douglas D H and Peucker T K 1973 Algorithms for the reduction of the number of points required to represent a digitized line or its caricature *Cartographica: the international journal for geographic information and geovisualization* **10(2)** 112-22

IOP Conf. Series: Materials Science and Engineering 1016 (2021) 012004 doi:10.1088/1757-899X/1016/1/012004

Methods of Determining Trajectory for Wheelchair with Manual Pushrims Drive

B Wieczorek

Institute of Machine Design, Poznan University of Technology,
3 Piotrowo street, Poznań 60-965, PL

*Email: bartosz.wieczorek@put.poznan.pl

Abstract. The article describes a method of operation of wheelchairs with differential control, as well as elaborates a developed method that allows visualising the wheelchair motion trajectory on the basis of reading of road or linear velocity measured independently on both left and right drive wheel. The Method chapter describes a complete proceeding algorithm, which allows determining the trajectory of wheelchair rear wheel axle's middle point by using only road value of rear wheels in regular intervals. The Measurement system chapter describes the used set of sensors, type of measured kinematic value, and method of transforming it into unit compliant with the method. The Results chapter shows exemplary values of measured wheelchair velocities and motion trajectory charts determined by transforming them.

1. Introduction

A wheelchair with manual drawdown drive is a part of the man-technical device anthropotechnical system that is responsible for the execution of locomotion function for a person with reduced mobility. A characteristic feature of this system is its propulsion method that consists of powering cycles [1] divided into propulsion phase, during which a propulsion force is generated and, return phase, during which hand returns to initial position [2]. The manual wheelchair drive system is characterised by its independence, which can translate to various drive torque values on both left and right drive wheel [3] that result from lack of symmetry in muscle strength of upper limbs [4, 5]. In addition, the difference in drive torque of drive wheels is also influenced by rolling resistance [6 – 9]. This is can be especially felt during the return phase because the system is not powered by impetus. The problem of increased motion resistances is also the result of utilising new technologies that support moving on a wheelchair [10]. Different values of drive torque on wheelchair's wheel translate into various velocity values for left and right drive wheel, which in consequence influences the wheelchair's trajectory. The familiarity of wheelchair's trajectory is particularly important during bench testing [11, 12] because it constitutes as feedback informing the patient sitting in an immobile position about distance travelled. In addition, as it is shown by research from the field of manual wheelchair propulsion biomechanics, the wheelchair's motion trajectory has an impact on other biomechanical parameters [13], such as centre of gravity location [14]. This impact results from dynamic changes in the position of human body during manoeuvring using a wheelchair [15, 16].

By observing the current trend consisting of development of manual drive systems by combining them with electrical drives [17], the familiarity of motion trajectory is also required during development of software controlling such hybrid drives. As a result, there occurs a necessity to elaborate a method of determining wheelchair's motion trajectory in iterative sequences using commonly available sensors. Additionally, the said sensors should also enable their implementation in the wheelchair's structure without excessively increasing its dimensions and mass.

2. Method

The developed method uses differential wheelchair control consisting of differentiating velocities independently of powered rear wheels v_L and v_P. According to the differential control rule, the R turning radius depends on the difference between velocity vectors of wheels (Fig. 1).

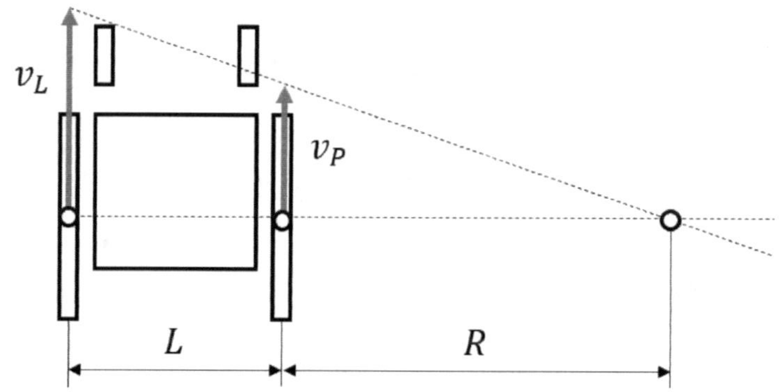

Figure 1. Determination of the R turning radius using geometric method on the basis of known velocity values of v_L left wheel, v_P right wheel, and L wheel spacing

In case of measuring kinematic parameters of a wheelchair, it is possible to measure an independent road for s_L left wheel and s_P right wheel within the same t time unit. Division of whole wheelchair motion into equal interval allows to divide the motion into trapezoids, the basis of which is equal to L wheel spacing and the sides of which are equal in length to road travelled by s_L left wheel and s_P right wheel (Fig. 2 A, B).

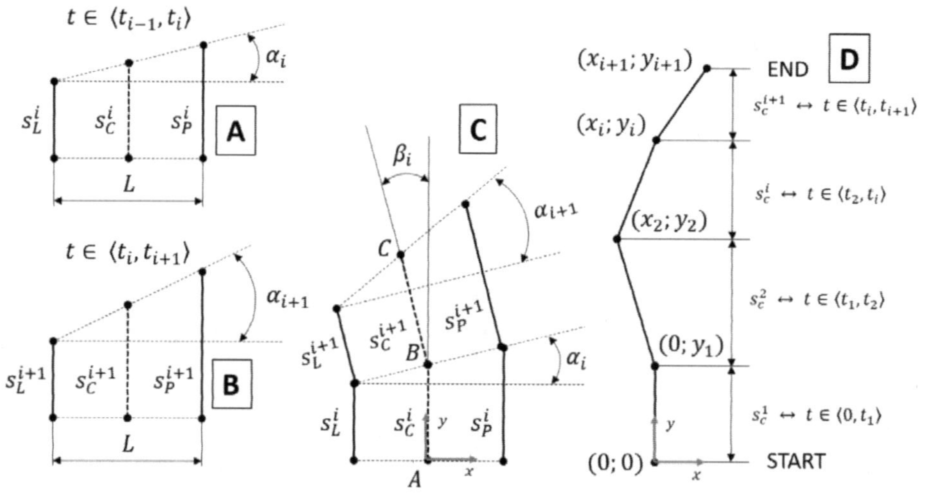

Figure 2. Diagram of distributing wheelchair motion trajectory into trapezoids (description in text)

According to the trapezoid method, first the whole motion should be divided into equal intervals for which the roads of s_L left wheel and s_P right wheel are known (Fig. 2 A, B). Next, the α inclination angle in relation to level of straight line connecting the ends of s_L left wheel and s_P right wheel should be calculated (1).

$$\alpha = atan\left(\frac{s_L^i - s_P^i}{L}\right) \quad (1)$$

The wheelchair's trajectory can be described as a change of position in space for a point constituting as the central point of wheelchair's rear drive wheel axle. Because of this, a path of s_C axle central point, which constitutes as mean of roads travelled by s_L left wheel and s_P right wheel, should be determined for each trapezoid describing the road at regular t intervals (2).

$$s_C^i = \frac{s_P^i - s_L^i}{2} \tag{2}$$

Determination of wheelchair's motion trajectory according to the trapezoid method consists of building whole wheelchair's road using specific trapezoid in compliance with several principles (Fig. 2 C, B). The first trapezoid framing the road of wheels and central axle point always has the *L* base parallel to the x axis describing the axis middle for time *t=0*, which constitutes as the start of coordinate system. Each subsequent trapezoid should be drawn in such a way that its L base will lie on the line inclined at the α angle and will connect the ends of s_L left wheel and s_P right wheel road from the previous propulsion cycle. In addition, the central points of drive wheel axle from the end of previous and beginning of new time interval should overlap as it is shown in case of point B (Fig. 2 C).

In order to determine wheelchair's trajectory, it is sufficient to determine the coordinates of drive wheel axle central point location for the end of each distributed propulsion intervals (Fig. 2 D). Coordinates x_0=0 and y_0=0 are adopted as the first point. Coordinates of the second point are x_1=0 and y_1=s_{C1}. Coordinates of each subsequent point are determined using equations (3, 4):

$$x_i = x_{i-1} + s_C^i \cdot \sin(\beta_i), \tag{3}$$

$$y_i = y_{i-1} + s_C^i \cdot \cos(\beta_i). \tag{4}$$

Where the β angle is described with equation that is the sum of all preceding α angles of the line connecting ends of roads for both s_L left wheel and s_P right wheel (5).

$$\beta_i = \beta_{i-1} + \alpha_{i-1} \tag{5}$$

3. Measurement system

The research was conducted using a measurement system consisting of two incremental encoders 1 connected with central unit 2 that transmits the measurement signal to in-house software (Fig. 3). The measurement system used Hohner 21-122-500 encoders with 500 impulse resolution. The angular velocity was determined using the ω_L left and ω_P right drive wheel. Next, the angular velocity was used to generate a histogram of wheelchair's velocity progress. With known linear velocity of wheelchair, the v(t) for each of independently powered wheels was determined for road travelled by s_L left wheel (6) and sP right wheel (7).

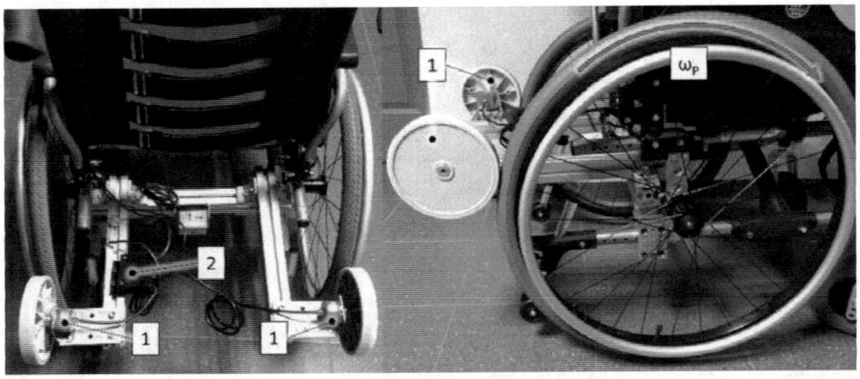

Figure 3. Wheelchair with test apparatus used to measure its motion kinematics, where: 1 - incremental encoder with 500 step resolution, central unit collecting measurement signal

$$s_L = \int_{t_{start}}^{t_{end}} v_L(t)dt = R\int_{t_{start}}^{t_{end}} \omega_L(t)dt \tag{6}$$

$$s_P = \int_{t_{start}}^{t_{end}} v_P(t)dt = R\int_{t_{start}}^{t_{end}} \omega_P(t)dt \tag{7}$$

where: s_L – road travelled by left wheelchair wheel within adopted time interval, : s_P – road travelled by right wheelchair wheel within adopted time interval, t_{start} – time of commencing a propulsion cycle, tend – time of concluding a propulsion cycle, v_L – left wheelchair wheel linear velocity measured with regular time interval, v_P – right wheelchair wheel linear velocity measured with regular time interval, ω_L – left wheelchair wheel angular velocity measured with regular time interval, ω_P – right wheelchair wheel angular velocity measured with regular time interval, R – radius of wheelchair's drive wheel.

4. Results

Method verification was conducted for laboratory measurements and measurements in real conditions. In case of laboratory measurements, the wheelchair was located on a stand equipped with traction rolls A (Fig. 4). Exemplary results of trajectories determined on the basis of known wheel velocities for bench testing were presented in Fig. 5 and 6.

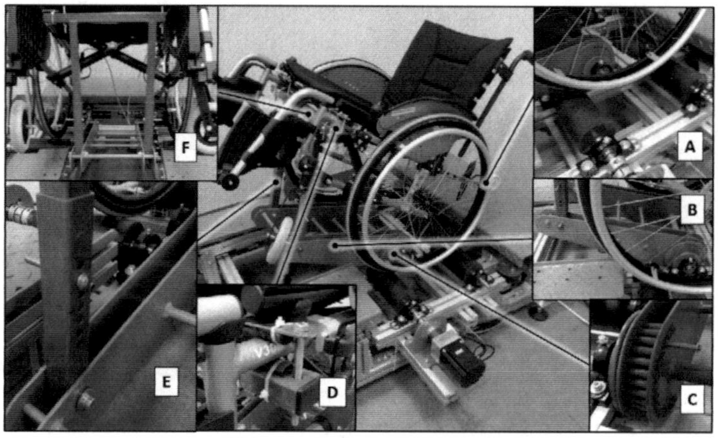

Figure 4. Examined wheelchair on a test stand with indicated wheelchair inclination system B-F and system of traction rolls receiving drive torque of a wheelchair's wheel.

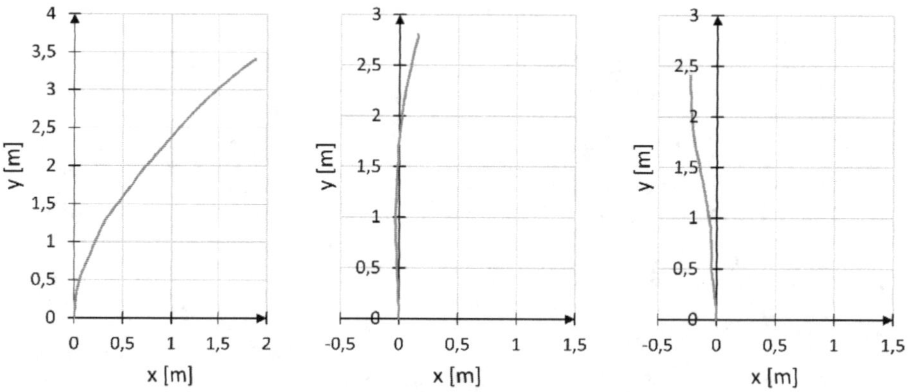

Figure 5. Exemplary wheelchair trajectory charts, where trajectory is determined during bench testing on the basis of velocity measurement

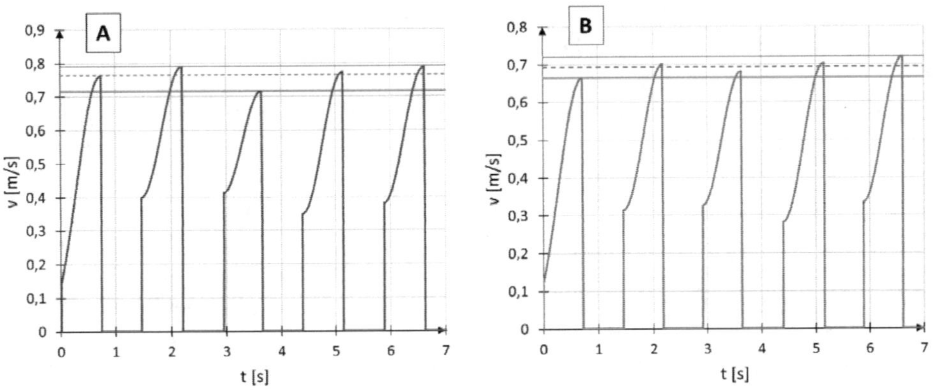

Figure 6. Velocity chart for left (A) and right (B) wheel measured on a test stand used to determine trajectory

In case of bench testing, the motion trajectory was determined only on the basis of road travelled by the wheelchair during propulsion phase and did not take into account road travelled during the phase of hand returning to initial position. This results from the construction of a stand, which does not allow to accumulate inertia force energy and as a result the wheelchair's wheel stops rotating when the hand returns to initial position during return phase.

The effectiveness of developed wheelchair trajectory determination method was also tested during propulsion the wheelchair in real conditions. Exemplary results of wheelchair's wheel velocity during travelling inside a building are shown in Fig. 7. Next, these results were used to visualise the wheelchair's motion trajectory (Fig. 8).

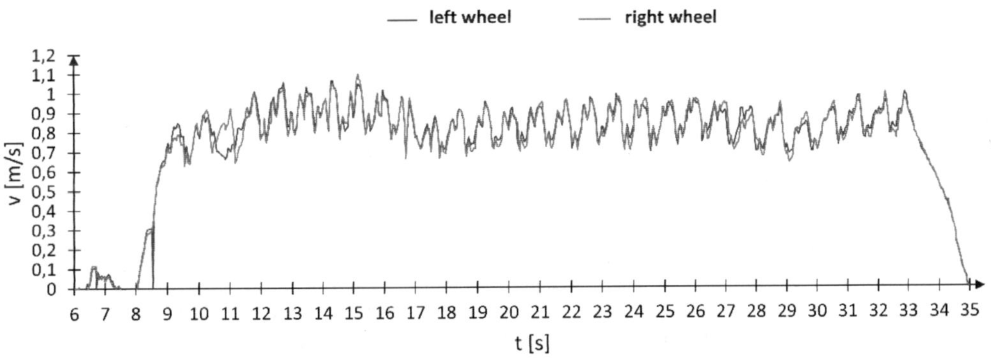

Figure 7. Chart of left and right wheel velocity measured during propulsion the wheelchair in real conditions and used to determine the wheelchair's motion trajectory

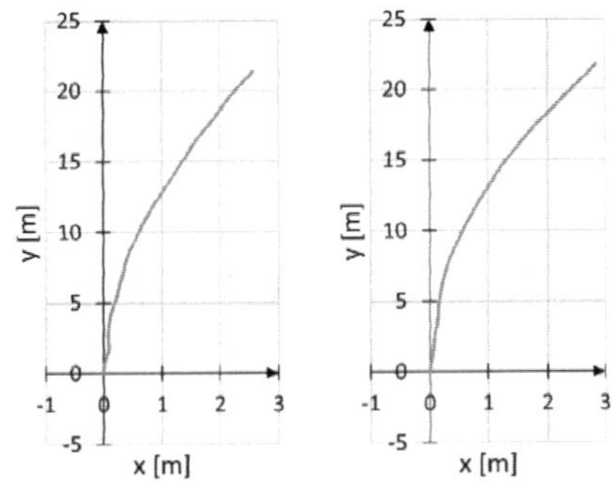

Figure 8. Exemplary trajectory charts for a wheelchair powered in real conditions

5. Summary

The developed method of determining wheelchair trajectory uses only measurement signal received from encoders. This translates into simple measurement system structure, which can be adopted in any wheelchair type and model equipped with manual drive. The described method proves itself as a tool that enables to draw trajectory in real time. Therefore, it is possible to create a feedback with user during operation in laboratory conditions. In addition, this method can constitute as a tool, which allows to analyse the wheelchair's motion kinematics and inspect the motion trajectory influence on biomechanical parameters [14] accompanying the combination of a human with wheelchair.

The accuracy of calculated trajectory depends on the sampling time and resolution of used encoders. The term sampling time mean time interval utilised to measure velocity or road travelled by drive wheel. Lowering the sampling time value results in greater the accuracy of generated trajectory. However, this translates into increase of iterations, which the program executing presented mathematical model must performed. As a result, a delay in the presented trajectory chart may occur. This phenomenon is especially undesired in case of presenting trajectory in real time.

The influence of encoder resolution on measurement accuracy results from the number of impulses that the encoder can generate during on full rotation of measurement wheel. In case of a too small encoder resolution, it is possible that a phenomenon, in which an identical amount of impulses generated by the encoder will be measured for several subsequent sampling times (iterations), will occur. When designing a test procedure, the sampling time should be appropriately selected depending not only on the encoder resolution, but also the envisioned wheelchair velocity. At high wheelchair velocities it is recommended to reduce the sampling time. However, the sampling time can be increased at low velocities.

Conclusion

The method described in the article has been verified during numerous measurement trials in laboratory and real conditions. The correctness of its operation has been confirmed during the said tests. Therefore, it became a useful tool in visualisation of road travelled by a wheelchair and other vehicles that are controlled differentially [18, 19]. Despite this, it is recommended to conduct further studies that will allow to determine the influence of used encoder's resolution and adopted sampling time as maximum wheelchair velocity ranges, for which the trajectory render accuracy is at a satisfying level.

Further plans include the development of this method, as well as its modification in order to enable trajectory visualisation in a three-dimensional system. For this purpose, it will be necessary to equip the measurement system with a gyroscope that will allow to measure the wheelchair inclination angle. Knowledge of this angle will enable the determination of changes in terrain height, on which the wheelchair moves. Another considered development direction for the measurement system is the method of coupling rotational velocity sensors with wheel. The utilised friction connection influences the deformation of wheelchair's tyre while leaving a connection through a strand-type transmission can provide a connection, which will not require adjustment during wheelchair's operation with various pressure in wheels or various wheelchair load conditions [20, 21].

References

[1] Van der Woude L H V Veeger H E J Rozendal R H and Sargeant A J 1989 Optimum cycle frequencies in hand-rim wheelchair propulsion *European journal of applied physiology and occupational physiology* **58(6)** 625-32

[2] Shimada S D Robertson R N Bonninger M L and Cooper R A 1998 Kinematic characterization of wheelchair propulsion *Journal of rehabilitation research and development* **35(2)** 210-8

[3] Kukla M Wieczorek B Warguła Ł and Berdychowski M 2019 An analytical model of the demand for propulsion torque during manual wheelchair propelling *Disability and Rehabilitation: Assistive Technology* 1-8.

[4] Morrow M M Rankin J W Neptune R R and Kaufman K R 2014 A comparison of static and dynamic optimization muscle force predictions during wheelchair propulsion *Journal of biomechanics* **47(14)** 3459-65

[5] Kukla M Wieczorek B and Warguła Ł 2018 Development of methods for performing the maximum voluntary contraction (MVC) test *MATEC Web of Conferences* **157** 05015

[6] Warguła Ł Wieczorek B and Kukla M 2019 The determination of the rolling resistance coefficient of objects equipped with the wheels and suspension system - results of pilot tests *MATEC Web of Conferences* **254** 01005 https://doi.org/10.1051/matecconf/201925401005

[7] Pałasz B Waluś K J and Warguła Ł 2019 The determination of the rolling resistance coefficient of a passenger vehicle with the use of roller test bench method *MATEC Web of Conferences* **254** 04007 https://doi.org/10.1051/matecconf/201925404007

[8] Pałasz B Waluś K J and Warguła Ł 2019 The determination of the rolling resistance coefficient of a passenger vehicle with the use of selected road tests methods *MATEC Web of Conferences* **254** 04006 https://doi.org/10.1051/matecconf/201925404006

[9] Sawicki P Waluś K J and Warguła Ł 2018 The comparative analysis of the rolling resistance coefficients depending on the type of surface – experimental research *Transport Means: Proceedings of the 22nd International Scientific Conference* 434-41

[10] Warguła Ł Kukla M and Wieczorek B 2020 The impact of wheelchairs driving support systems on the rolling resistance coefficient *IOP Conf. Series: Materials Science and Engineering* **776** 012076 https://doi:10.1088/1757-899X/776/1/012076

[11] Wieczorek B and Warguła Ł 2019 Problems of dynamometer construction for wheelchairs and simulation of push motion *MATEC Web of Conferences* **254** 01006

[12] Kukla M Wieczorek B Warguła Ł and Górecki J 2019 The determination of the parameters of wheelchair driving with the use of a test bench *Autobusy: technika, eksploatacja, systemy transportowe* **20**

[13] Wieczorek B and Kukla M 2020 Biomechanical Relationships Between Manual Wheelchair Steering and the Position of the Human Body's Center of Gravity *Journal of Biomechanical Engineering* **142(8)**

[14] Wieczorek B and Kukla M 2019 Effects of the performance parameters of a wheelchair on the changes in the position of the centre of gravity of the human body in dynamic condition *PloS one* **14(12)** e0226013

[15] Wieczorek B Kukla M and Warguła Ł 2020 Methods for measuring the position of the center of gravity of an anthropotechnic human-wheelchair system in dynamic conditions *Materials Science and Engineering Conference Series* **776(1)** 012062

[16] Wieczorek B Górecki J Kukla M and Wojtokowiak D 2017 The analytical method of determining the center of gravity of a person propelling a manual wheelchair *Procedia Engineering* **177** 405-10

[17] Wieczorek B Warguła Ł and Rybarczyk D 2020 Impact of a hybrid assisted wheelchair propulsion system on motion kinematics during climbing up a slope *Applied Sciences* **10(3)** 1025

[18] Rybarczyk D and Milecki A 2020 Electrohydraulic Drive with a Flow Valve Controlled by a Permanent Magnet Synchronous Motor *Transactions of FAMENA* **44(2)** 0-0

[19] Waluś K J Warguła Ł Krawiec P and Adamiec J M 2017 The impact of the modernization of the injection-ignition on the parameters of motion of the motorcycle *Procedia Engineering* **177** 393-8

[20] Krawiec P Różański L Czarnecka-Komorowska D and Warguła Ł 2020 Evaluation of the Thermal Stability and Surface Characteristics of Thermoplastic Polyurethane V-Belt *Materials* **13** 1502

[21] Krawiec P Pajtášová M Meler F and Warguła Ł 2020 Testing functional features of V-belt transmissions *IOP Conf. Series: Materials Science and Engineering* **776** 012008 https://doi:10.1088/1757-899X/776/1/012008

Acknowledgments

The research was made as part of the Lider VII project “Study of the biomechanics of manually propelled wheelchair for innovative manual and hybrid drives” (LIDER/7/0025/L-7/15/2016) financed by the National Centre for Research and Development.

IOP Conf. Series: Materials Science and Engineering 1016 (2021) 012005 doi:10.1088/1757-899X/1016/1/012005

Lesson Learned from Semester of Online Teaching of Automation Using Simulators

I Ostrowski, J Możaryn, J Wang and D Cukier

Faculty of Mechatronics, Warsaw University of Technology, Warsaw, Poland

*Email: igor.ostrowski@pw.edu.pl

Abstract. The year 2020 has taught us a very important lesson about communication between university lecturers and students in laboratories. This paper is showing conclusions after a semester of remote personal contact with students and laboratory equipment. In normal cases authors of this paper providing their lessons in a laboratory equipped with industry effectors, Robots, Pneumatic equipment, and PLC's. It was replaced with simulators, because of the lockdown due to the coronavirus pandemic. In the paper, we describe the on-line didactic process conducted with simulators such as Factory IO; Festo FluidSim; Roboguide.

1. Introduction

In the year 2020, the COVID-19 pandemic situation forced the standard university lectures to be conducted remotely. However, there are areas where it would seem that replacing the student's personal presence, like in the laboratory, it will have a negative effect in the form of low teaching efficiency. The belief that the learning outcomes are lower may come from the deeply rooted opinion that personal participation in classes brings the most effective didactic effect. This claim is based on an often-cited work by Edgar Dale [1]. The results of Dale's research, however, are usually overinterpreted, which leads to the erroneous definition of human perceptual abilities in distance learning [2]. Although the software environments presented in our study have so far served as a supplement to the normal curriculum, in 2020 they completely replaced the teaching work with students in the laboratory. In the teaching of machine operation and control engineering, the simulators and on-line methods are fully justified as proven and effective tools [3]. In teaching programming languages and computer technologies, as well as soft skills, distance learning methods are successful and have been confirmed by the long-standing operation of many learning platforms [8],[9,][10]. In teaching elementary skills in the field of automation and robotics, the last years' experience of forcing online teaching of what seems to have to be taught through face-to-face contact with the teacher and equipment brings an interesting lesson. All the presented here cases are based on the experience gained by authors at the Institute of Automation and Robotics, Warsaw University of Technology.

2. Software description and use cases

2.1. Factory IO

Factory I/O [11] is an environment for 3D simulation of production lines for the design and testing of industrial process automation systems [6], [7]. The software is designed as an easy-to-use program that allows you to quickly build a virtual factory using parts from popular manufacturers. Factory I/O works with simulators of PLC controllers from leading automation manufacturers, including Siemens, Allen-Bradley, among others. Factory I/O also includes many scenes with typical industrial installations.

IOP Conf. Series: Materials Science and Engineering 1016 (2021) 012005 doi:10.1088/1757-899X/1016/1/012005

The software package offers a collection of components that are used in current production lines. Such elements include conveyor belt modules, transport roller systems, pallet lifts, CNC stations cooperating with robots, etc. The simulated machine park is supplemented by a set of various sensors, thanks to which it is possible to build and set-up a sensory system for the production processes.

THE factory I/O environment works very well as a training platform for PLC, SoftPLC, Modbus TCP/IP, etc. In the didactic process at the Institute of Automation and Robotics, Warsaw University of Technology, Factory IO software was used in the preparation of theses on the machine automation system design. The aim of one of such work was to design an automated production process with a safety system. A simulation of the production and safety system was carried out using the virtual PLC Siemens SIMATIC S7-1200 and the TIA Portal software (Figure. 1).

The project has been divided into stages that showed the details related to the design, programming, and simulation of the selected production line, including the transport line for processed elements, plastic elements processing lines (Figure. 2), and the carton packaging and storage line (Figure. 1). The implementation of all these elements made it possible to implement the project of a simulated production line with a safety system and to propose and conduct a simulated safety audit for all machines during the operation of the finished line. Audit objectives were adapted to the existing machines, which made it possible to analyze the operation of the safety system. The thesis also included a description of single and comprehensive production line tests and their documentation. Additionally, the project was expanded with a video showing the overall operation of the lines, individual machines, and the detailed work of the safety system with violations.

Due to the difficult contact with the student caused by lockdown, most consultations were carried out remotely. The work was done during the summer semester. The final defense was also carried out remotely, following the rules adopted at the Warsaw University of Technology, developed for the university's work in the event of the COVID-19 pandemic.

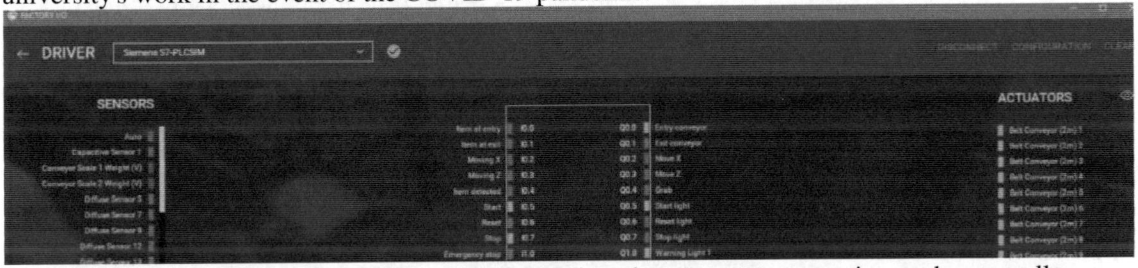

Figure 1. Factory I / O window with information about correct connection to the controller

Figure 2. The carton packaging and storage line, during safety system tests

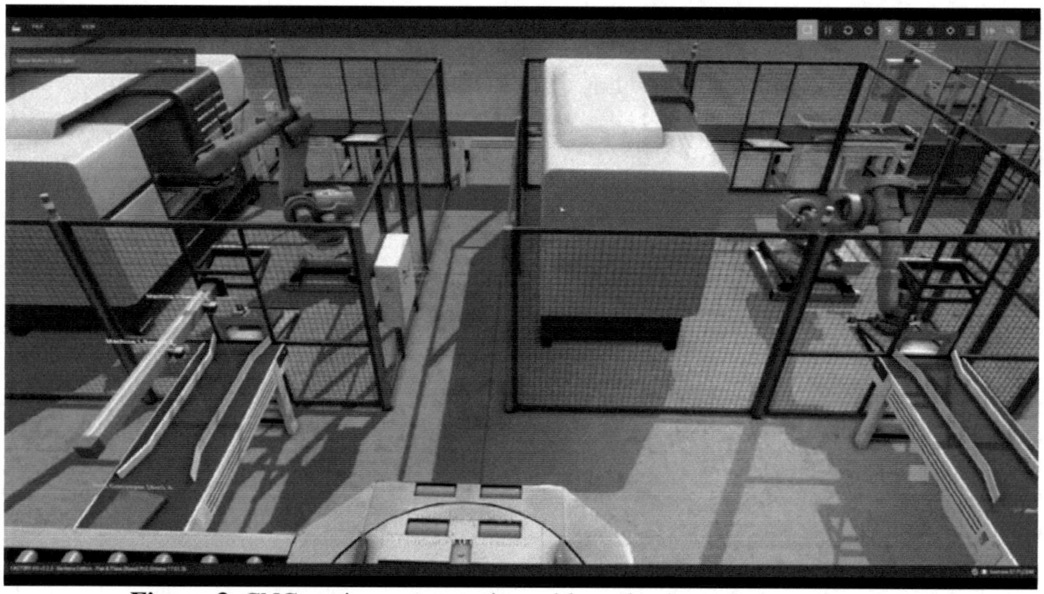

Figure 3. CNC stations cooperating with packaging robot manipulators

2.2. FluidSim

In the Institute of Automatic Control and Robotics, the program FluidSim [12] was used to learn basic pneumatic, hydraulic and electrotechnical connections (Figure. 4) in the form of sharing the instructor's window, who discussed the schematics on-line. Moreover, the FluidSim environment was used to describe the functions of the PLC controller in automation systems. Diagrams of fluid and pneumatic automation systems simulated in the program are in accordance with DIN ISO 1219. The program allows users to generate technical documentation on the basis of simulated systems.

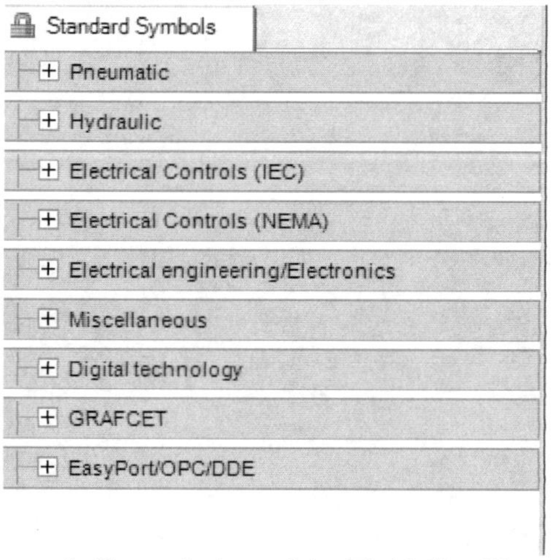

Figure 4. General view of the Fluid Sim libraries

According to the manufacturer's description, the program is a good self-study tool for the student. On the FluidSim website, one can find that "*FluidSIM 5 is a comprehensive software for the creation, simulation, instruction and study of electro-pneumatic, electro-hydraulic, digital and electronic circuits. All of the program functions interact smoothly, combining different media forms and sources of knowledge in an easily accessible fashion. FluidSIM unites an intuitive circuit diagram editor with detailed descriptions of all components, component photos, sectional view animations, and video sequences. As a result, FluidSIM is perfect not only for use in lessons but also for the preparation thereof and as a self-study program.*".

The Microsoft Teams, considered as leading distance learning tool [4], [5], was used to connect students and lecturers. As part of the lectures, students were asked to make connections that performed specific discrete functions (Figure. 5). The tasks were transferred in text form, with a specified deadline for completion of works. The students' work was presented in the form of screenshots, program files, and video recordings of the virtual layout. A valuable experience is the ability to learn how to use the program remotely and the statement that the very operation of the program causes the students to ask questions about the phenomena occurring in a real facility compared to the simulator. The inability to conduct any other than a virtual experiment causes that they accept the answer without checking the real mechanism. In case of difficulties in understanding the task or its execution, the tutor can take control of the student's desktop, making corrections and giving instructions.

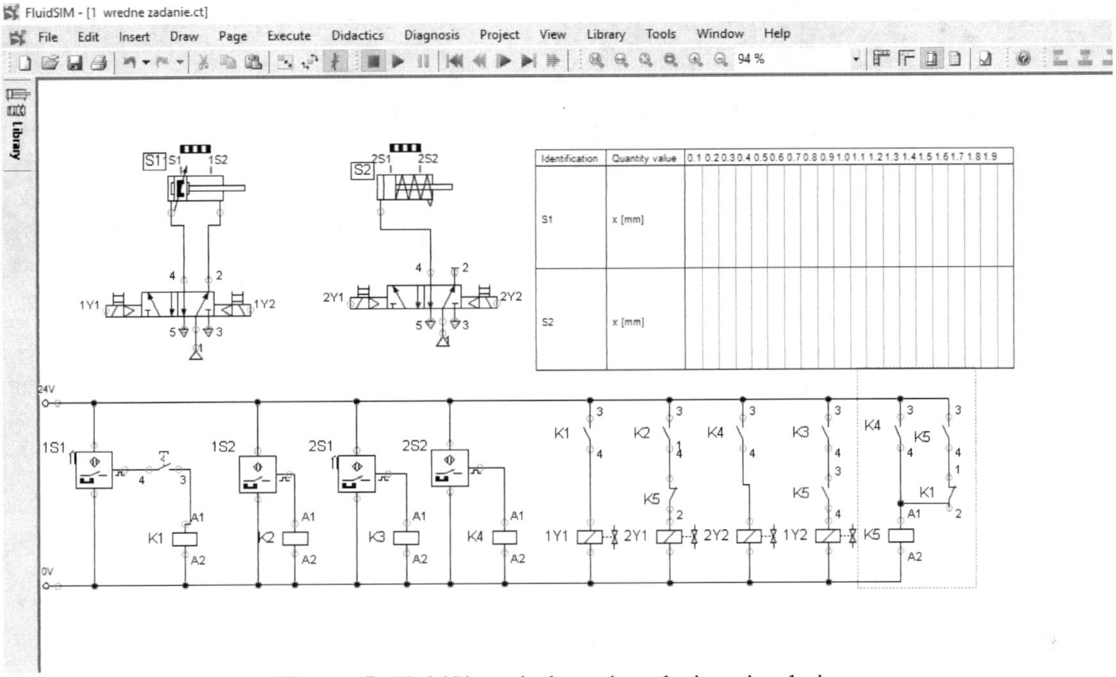

Figure 5. FluidSim window view during simulation

The use of Fluid Sim during the lockdown due to COVID-19 pandemic allowed to teach students symbols compliant with the standards, rules of creating technical documentation, basic physical dependencies, as well as elementary combinational and sequential systems.

2.3. Roboguide

Roboguide is an offline FANUC robotics simulation software developed by FANUC. It contains varieties of FANUC robot models and controllers which could easily simulate the tasks in work cells. Roboguide allows the users to fully simulate the tasks in a 3D virtual environment to minimize unexpected programming issues without physical robots. Together with the simulated teach-pendant and diversity option of tools, Roboguide could satisfy the wide range of robot applications in the industries [13].

As the pandemic limited us for lecturing the student to manage the knowledge of robots in the lab, Roboguide became a highly practical tool for teaching. The students could make the simulation of the robots and manage the knowledge after remotely teaching and tutorials.

From the teaching perspective, setting tool frames and user frames of robots are the fundamental skills that students should manage. The part of the teaching activities was done by watching a series of videos we prepared in our lab and the remote lecture basis on the physical FANUC robot (Figure.6) and the other was performed by Roboguide. In the video, we demonstrate to the students how to use the teach-pedant of FANUC robot and how exactly setting the tool frames and user frames.

IOP Conf. Series: Materials Science and Engineering 1016 (2021) 012005 doi:10.1088/1757-899X/1016/1/012005

Figure 6. Video made for teaching students how to set the tool frame

As simulation software, Roboguide's interface and some functions are different from the physical robots. Therefore, the training of operating the software is necessary and could make the students understand better the knowledge again. After the remote lectures and watching the videos, the interests of students in the robots are aroused, and instruction and demonstration of using Roboguide to practice were arranged (Figure. 7).

Roboguide is commercial professional software, full version must be purchased. Fortunately, the students could install the academic version which is enough for the practice. In the remote lectures and tutorials prepared for basic Roboguide usage, we perform the teaching actions in Table 1.

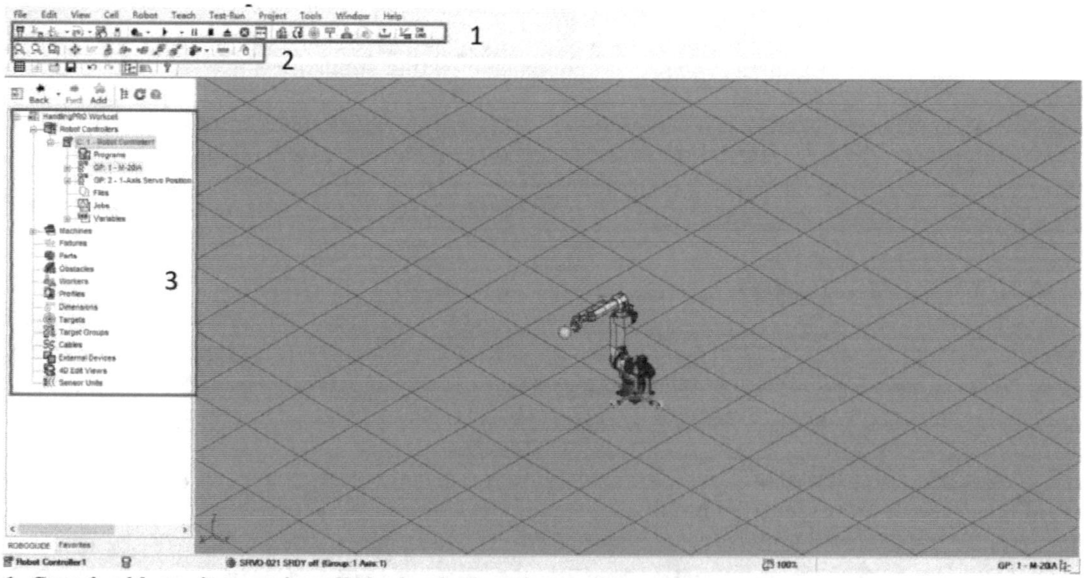

1. Standard bar - it contains all the basic functions of the robot.
2. View bar - camera operation and distance measurement
3. Cell Browser - program structure viewer

Figure 7. The instruction we prepared for students to understand the functions of Roboguide

Table 1. Basic Roboguide teaching arrangement

	Teaching steps	The skills students learnt
1	Roboguide introduction	Students learn how to use virtual teach pendant and the difference from the physical robot
2	Roboguide setting tool and user frames	The students choose the proper tools and workspace and correctly set the tool and user frames
3	Roboguide pattern drawing simulation	The students could use make basic programs and perform certain task in Roboguide

When the Roboguide remote demonstration and tutorial are completed, the tasks are assigned to the students. The tasks are drawing different patterns and welding a part using Roboguide. The students shall send back the .frw and .AVI exported from Roboguide to the lectures for checking the correctness. Roboguide provides several different exported format such as .AVI or .frw for easier checking if the task was performed smoothly or need some corrections which are friendly for students and lectures (Figure.8).

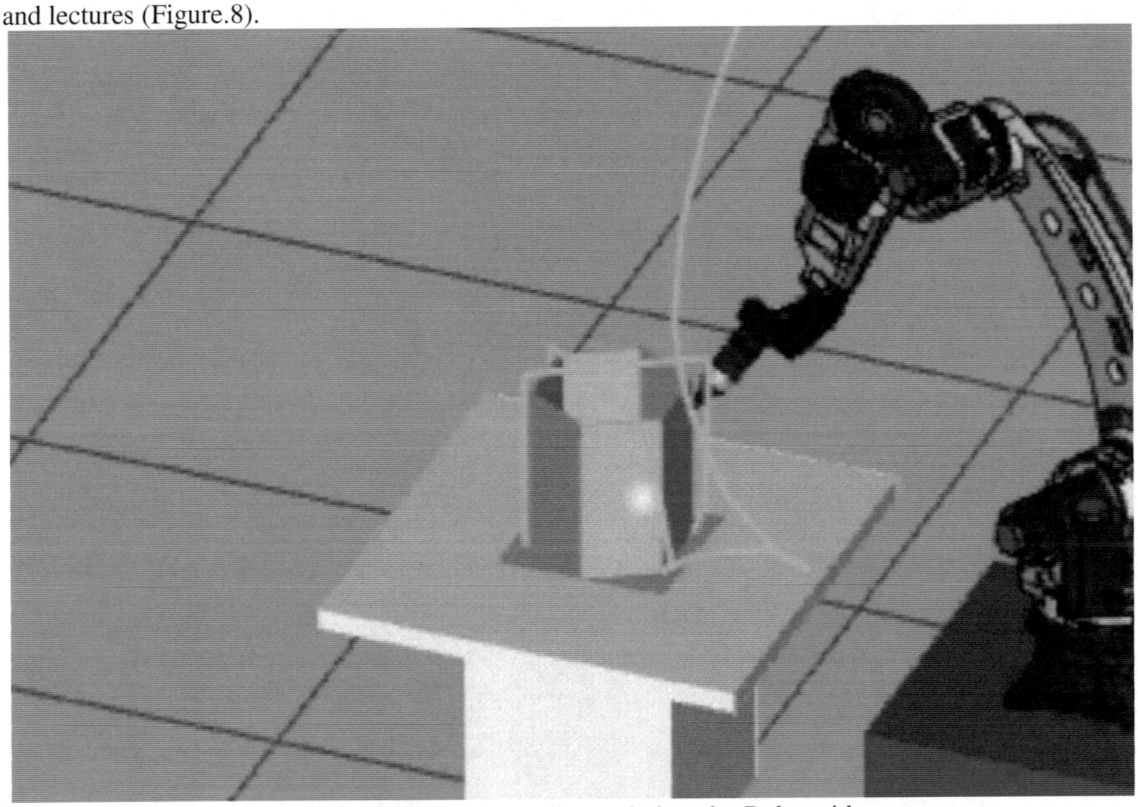

Figure 8. Students welding task done by Roboguide

In summarize, despite the pandemic limited actual classes and labs, the students could keep on learning in remote way. Roboguide works not as FANUC robot simulator but as a great remote teaching tool for students and lecturers without physical contact.

Conclusion

Remote work requires the preparation of separate tools that were not in such extended usage before the pandemic or were used for other purposes. Distance learning methods took time to develop. Our vision of classes conducted with the use of simulators has been developed with videos from laboratories. Classes in the second wave of the pandemic are prepared much more carefully, their effectiveness will probably also be higher. We assume that the next step will be on-line video from the studio, in which the teacher will present the exercise and the recording of these films will be accessed by students.

The lecturers of the described classes agree in the conclusion that the effectiveness and comfort of expanding knowledge are greater in the classes conducted in person. Moreover, it does not cause a feeling of isolation in students and the teacher. However, we would like to point out that the methods of digital participation in classes allow for effective transfer of knowledge and can be highly useful in educating people who, for various reasons, cannot come to the labs and classrooms. Lessons learned from the difficult year 2020 will allow us to prepare better teaching materials for distance learning, if necessary. These materials can also be used as complementary to the standard live teaching process with the possibility of self-study.

References

[1] Dale E 1969 *Audio-Visual Methods in Teaching 3rd Edition* (Holt, Rinehart, and Winston)
[2] Subramony D Molenda M Betrus A and Thalheimer W 2014 Timeline of the Mythical Retention Chart and Corrupted Dale's Cone *Educational Technology* **54(6)** 31-4
[3] Ostrowski I and Maslowski A 2019 Simulator effectiveness test for e-training with the use DROMADER tele-operated vehicle *Acta Imeko* **8(4)** 54-61
[4] Davidson P Long E Molnar A Chui R T M and Ting C Y 2018 Ms teams and google classroom: preliminary qualitative comparisons and user feedback *5th Pre-University Conference* 107
[5] Hai-Jew Shalin 2020 Evaluating "MS Teams" for Teaching and Learning *C2C Digital Magazine* **1(13)** 7
[6] Riera B and Vigário B 2017 HOME I/O and FACTORY I/O: a virtual house and a virtual plant for control education, *IFAC-Papers OnLine* **50(1)** 9144-9
[7] Philippot A et al 2017 HOME I/O and FACTORY I/O: 2 Pieces of innovative PO simulation software for automation education *27th EAEEIE Annual Conference (EAEEIE)* 1-6
[8] http://www.mooc.org
[9] http://www.udemy.com
[10] http://www.udacity.com
[11] http://factoryio.com
[12] http://www.art-systems.de/www/site/en/fluidsim
[13] https://www.fanuc.eu/pl/en/robots/accessories/roboguide

Development and Modeling of Remote Laboratory Works for Engineering Education

B B Sus[1]*, S P Zagorodnyuk[2], O S Bauzha[2] and A Kozinetz[1]

[1]Department of Nanophysics of Condensed Matter,
Taras Shevchenko National University of Kyiv, Kyiv 01033, Ukraine
[2]Department of Computer Engineering,
Taras Shevchenko National University of Kyiv, Kyiv 01033, Ukraine

*Email: bnsuse@gmail.com

Abstract. The basic principles of finite-state machine synthesis and simulation on Proteus CAD and VHDL are discussed. The main steps of designing an electronic circuit of the control and operating finite-state machines that perform the calculation of a given algebraic expression were described. The main steps of consistent performance of a complex task for the development of cross-cutting knowledge and skills, which must be performed by the student in the framework of the normative discipline ware analysed in detail. It is shown that the developed cycle of laboratory works can be successfully implemented in various training courses.

1. Introduction

The application and effectiveness of distance and virtual works in the educational process of engineering specialties have some new features and benefits such as clarity and convenience of modeling processes and phenomena, closer mutual integration of disciplines, standardized curriculum, the formalized procedure for assessing technical tasks, and flexible system of sufficient formation of assignments. It turns out that the solution to this class of technical problems is almost impossible without the widespread use of specialized computer-aided design systems.

One such effective computer-aided design system, which allows the synthesis, circuit modeling, and simulation of electronic circuits in real-time, is the software package Proteus Design Suite.

The Proteus simulator library has a large number of ready-made electronic components: power supplies, generators, signal analyzers (oscilloscope, voltmeter, ammeter, frequency analyzer), indicators for monitoring the parameters of the circuit in real-time (probes), discrete components, including resistors, capacitors, transformers, sensors and others. Simulation of the real behavior of electronic components and circuits based on the use of a specialized modeling language SPICE (Simulation Program with Integrated Circuit Emphasis). As a result of the evolution of software emulators, the SPICE language has become the major language for most programs that simulate the behavior of electronic components.

The software package Proteus VSM allows the student to simulate both analog and digital circuits, their combination, as well as simulation of complex sensory instruments and electronic systems on microcontrollers [1, 2]. The main advantage of the Proteus software simulator is the presence of libraries that include a large number of microcontrollers, microprocessors and logic elements. Using the Proteus environment allows you to visualize the functioning of the nodes of the computer system at the micro-level. In particular, an laboratory evaluation of the efficiency of using the Proteus virtual environment was discussed in [3]. The simulation of a digital device for measuring alternating current, voltage and power has been performed in [4]. Hardware implementations of the electric brushless motor driver circuit with three stator windings, as well as modeling of the driver circuit in the Proteus CAD have been discussed in [5] and visual 3-d modeling is described in [6]. In particular, performance evaluation of microcontroller driver implemented in the circuit was made by the means of well-known popular C-language model library. The article demonstrates that the logic of the microcontroller program can be constructed, debugged and compiled on an independent computer in a programming environment for any model of microcontroller available in Proteus, and then the binary file of the compiled program can be downloaded directly to the simulation unit of the selected microcontroller model.

Another alternative, but very effective and powerful way to form and test the logic of the electronic circuit is to describe the operation of the new synthesized device in a specialized language of equipment description. There are many hardware description languages, but the most popular are VHDL and System Verilog. The language VHDL (Eng. VHSIC - Very high speed integrated circuits Hardware Description Language) is the basic language in the development of devices for modern computer systems. For example, the problem of designing and modeling an electric motor driver, similar to [5], has also been successfully solved using the VHDL language in solar cell rotation systems [7] and the vehicle automation system [8].

VHDL is a tool for describing mostly digital systems, but there is a modification of this language called "VHDL AMS" (Analog and Mixed Signals), which allows us to describe both analog and combined digital-analog circuits [9]. Typically, the analog part of the circuit get the analog signal of the sensor set and digitizes it using an ADC. Then the digital signal is processed by the main digital section of the mixed circuit. Such the mechanism used by medical diagnostic equipment, for the design of which the VHDL AMS language is best fitted [10].

Composing an electronic circuit in hardware description languages opens the developer the opportunity to create circuits of arbitrary complexity, including developing new processors with a limited set of instructions [11], or new specific instructions [12]. A designed, debugged and successfully compiled electronic circuit in a VHDL or System Verilog application development environment can further be simulated using other automated design systems, such as ModelSIM or Simulink [13]. Also, the binary file of the compiled VHDL / Verilog program can also be downloaded to the programmable logic integrated circuit (Programmable Logic Device, PLD) - a specialized minicomputer, the logic of which is not determined during the manufacture of the device, but determined by the developer using programming languages equipment.

Modern PLD controllers have a wide range of additional hardware-implemented auxiliary components, including signal generators and analyzers, Digital Signal Processor, logic elements, addition and multiplication devices, input, output and information storage devices [14]. Similar to the library components of the Proteus CAD, optional components of the PLD minicomputer can be handled by the developer together with the new electronic circuit. This type of modern PLD device is called a "Field-programmable gate array" (FPGA). The developer can easily convert the FPGA controller into an experimentally functioning instance of its integrated circuit and can then supply the input of this controller with real electrical signals and receive the actual generated signals at the output of the controller [15]. Therefore, the logic of the FPGA-controller can be said that one part of it is implemented during the manufacture of the controller and can not be redefined by the developer, and the other part can be designed by the inventor, who, conversely, can repeatedly change, supplement and expand it.

The article talks about the procedure of designing a complex electronic circuit using Mill finite state machines using the Proteus automatic design system, as well as the implementation of this circuit in VHDL and execution of this circuit on an FPGA-controller Altera Cyclone-II. Similar problems in the designing of digital circuits based on Moore's and Mealy finite-state machines are fully in line with global trends and are actively solved by research engineers. [16].

2. Formulation of the problem.

During the research, a series of laboratory works were developed. It can be used in various training courses: "Digital Circuitry and Electronics", "Intelligent Systems, Control and Automation", "Computer Logic" and "Microprocessor Engineering", and therefore the development took into account the interdisciplinary approach. The study of these courses provides the student the opportunity to gain fundamental knowledge and experience in the development and synthesis of electronic circuits with an assigned functionality. These disciplines give students the opportunity to consistently expand and increase the valuable properties of new devices. Consequently, in the course of "Digital Circuitry and Electronics" simple logic gates and combinational circuits and triggers were considered in detail. In the course of "Intelligent Systems, Control and Automation" complex combinational circuits and digital finite-state machines are already considered with the elements of the previous course that are already used as ready modules. The course of "Computer Logic" analyzes the structure of the operating machine and the logic of its functioning. The course of "Microprocessor Engineering" examines the actual microcontrollers with the peculiarities of their architecture.

Mastering these courses requires a clear sequence of learning and also involves the use of modern CAD-systems for the design and simulation of devices synthesized by students. The article demonstrates in detail the main steps of consistent implementation of a complex task that must be performed by the student in the framework of the discipline "Computer Logic".

The task for the student is to develop an operating scheme of a computing device that implements a certain computational operation. This operation may consist of the following arithmetic operations: addition, subtraction, multiplication, division, root extraction, multiplication and division by 2 (offset left and right), synthesis and construction of the functional diagram of the operating and control machines, which together provide control of the computing device. During the laboratory work, the student is limited in the choice of the element base of logical devices, types of triggers as memory cells, as well as algorithms for operations of multiplication, division, and root extraction. The large variety of elements is excessive, excessive, and significantly reduces the versatility of the synthesized device. It also requires the student to perform their assignment, which differs from the tasks of other students. A total of 1024 modifications of tasks have been developed for this work.

During laboratory work, it is very important to optimize the schematic diagram, reduce the number of logical elements, since it forms the economic efficiency of the subsequent hardware-implemented circuit.

Assume that the computational operation to be implemented is the calculation of the expression 4C 2AB. Use the Mealy finite-state machine as the controller, and use the D-flip-flops as the memory cells. Also, the student should implement a multiplication algorithm: "Multiplication from the lower digits of the multiplier and the shift of the sum of partial terms to the right."

The multiplication operation is performed in direct codes. Sign and main digits are processed independently. To determine the sign of the result of multiplication, XOR operation on the sign bits of the multiplied numbers is performed.

Let the multiplied Y (Y= 0, y1 , y2 ,…, yn ; yi be {0,1}) and the factor X (X =0, x1 , x2 ,…, xn ; xi be {0,1}). Then the result of multiplication of the absolute values of Y та X can be expressed as:

$$Z = YX = Yx_1 2^{-1} + Yx_2 2^{-2} + \ldots + Yx_n 2^{-n}$$

Multiplication of two numbers Y and X can be realized by performing a cyclic process, the nature of which depends on the specific form of expression. Each multiplication cycle consists of adding the next partial product, which is the product of a multiplier multiplied by one digit, to the sum of the partial products. We present the multiplication as:

$$Z = YX = ((\ldots((0+Yx_n)\, 2^{-1} + Yx_{n-1})\, 2^{-1} + \ldots + Yx_i)\, 2^{-1} + \ldots + Yx_1)2^{-1}$$

From the obtained expression it is seen that if we denote by $Z_i = (Z_{i-1} + Yx_{n-i+1})*2^{-1}$ where i is {1..n}, then then we can find the value of the result of multiplication by iteration. Initial values of i = 1, $Z_0 = 0$. At the end of n iterations $Z_n = Z = YX$.

IOP Conf. Series: Materials Science and Engineering 1016 (2021) 012006 doi:10.1088/1757-899X/1016/1/012006

In this method, multiplication is performed from the lower digits of the multiplier, the sum of partial results of multiplication is shifted to the right, and the multiplied remains fixed. The operating scheme of the multiplication machine according to the first algorithm is presented in Fig. 1. The multiplied (Y) is expressed in the lower bits of the register RG3, the multiplier (X) is written in the register RG2. The result of the multiplication (Z) will be obtained in the registers RG1 (senior digits) and RG2 (junior digits). Before multiplication, the register RG1 is set to zero. Counting the number of multiplication cycles is provided by CT counters, according to which its bit rate q, $2q \geq n$ is chosen.

The algorithm for obtaining the result in this method is as follows:

1. Enter the multiplied (Y) in the lower digits of the register RG3.
2. Enter the factor (X) in the register RG2.
3. The contents of the adder (register RG1) are reset.
4. Add to the register RG1 the contents of RG3 (when the least significant bit RG2 is 1), or 0 (when the least significant bit RG2 is 0). The bit matrix AND is used.
5. Move to the right (division by 2) registers RG1 and RG2. During the offset, the least significant bit of the register RG1 crosses to the most significant bit of the register RG2, and the least significant bit of the register RG2 is lost.
6. Repeat steps 4 and 5 (n-1) more times.

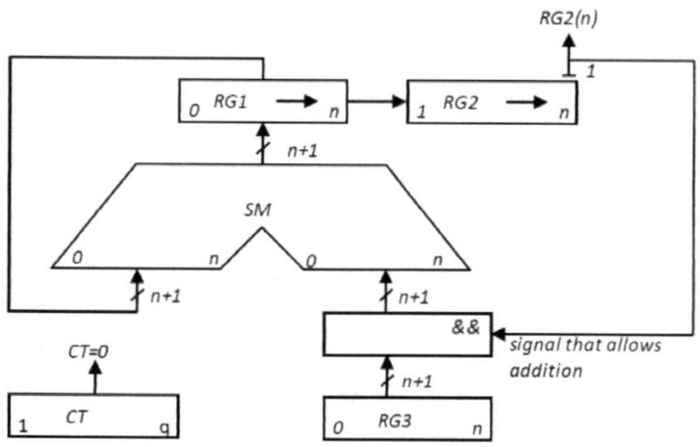

Figure 1. Operating diagram of the multiplication device.

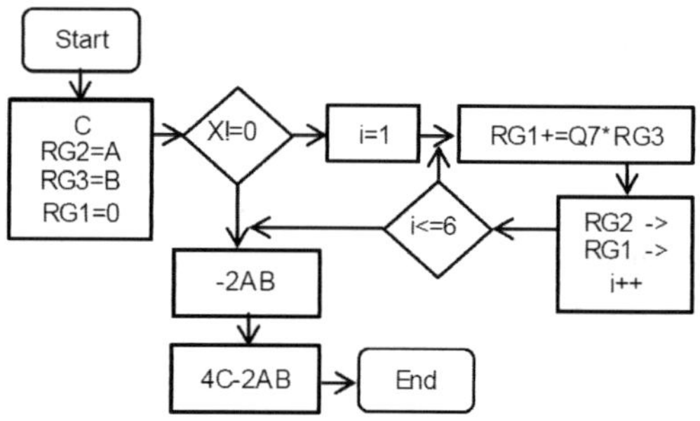

Figure 2. Essential scheme of the algorithm

A constructed table of coding of vertices and the graph-scheme of transitions of the datapath is presented in fig.3.

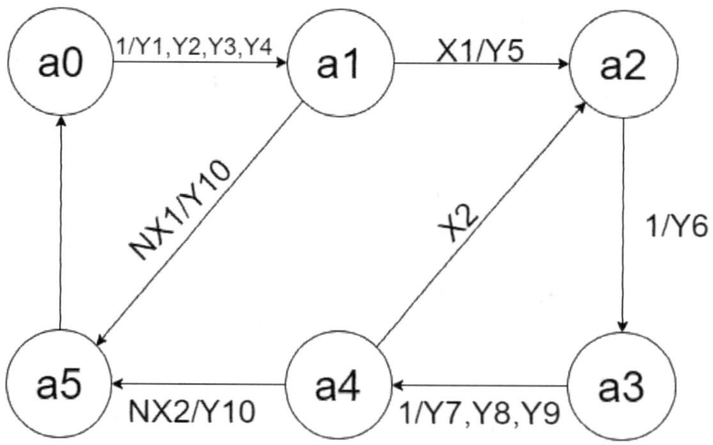

Figure 3. Graph diagram of the transitions of the datapath.

IOP Conf. Series: Materials Science and Engineering 1016 (2021) 012006 doi:10.1088/1757-899X/1016/1/012006

Table 1. Vertex coding table.

Code	Content	Notes
Y1	C	Enter C
Y2	RG2=A	Write in RG2=A
Y3	RG3=B	Write in RG3=B
Y4	RG1=0	Fill RG1 with zero
Y5	i=1	Counter is set to zero
Y6	RG1+=Q7*RG3	If the las bit RG2[Q7]=0, then write the zeros into RG1, if 1, then write the bits from RG3 into RG1
Y7	RG2->	Shift RG2
Y8	RG1->	Shift RG1
Y9	i++	Counter increment
Y10	-2AB	Convert AB into the supplementary code
Y11	4C-2AB	Calculate 4C-2AB
X1	X!=0	Check the multiplier by 0
X2	i<=6	Check the counter at the end of the multiplication

Encoding of the state of the machine is presented it table 2:

Table 2. Encoding of the state of the finite-state machine.

Status	Code	Decimal notation
a0	101	5
a1	010	2
a2	100	1
a3	110	3
a4	001	4
a5	000	0

We obtain a structural table of transitions of the datapath, from which we build a system of equations of outputs and a system of equations of transitions of the finite-state control machine.

Table 3. Structural table of transitions-outputs of the Mealy finite-state machine.

S	E	X	Y	Ks	Ke	Ф3
a0	a1	1	Y1,Y2,Y3,Y4	101	010	D2
a1	a2	X1	Y5	010	100	D1
	a5	NX1	Y10	010	000	-
a2	a3	1	Y5	100	110	D1D2
a3	a4	1	Y7,Y8,Y9	110	001	D3
a4	a2	X2	-	001	100	D1
	a5	NX2	Y10	001	000	-
a5	a0	1	Y11	000	101	D1D3

System of transition equations:

$D1 = a_1x_1 \vee a_2 \vee a_4x_2 \vee a_5$

$D2 = a_0 \vee a_2$

$D3 = a_3 \vee a_5$

System of output equations:

$Y1=Y2=Y3=Y4=a_0$

$Y5=a_1x_1$

$Y6=a_2$

$Y7=Y8=Y9=a_3$

$Y10=a_1\bar{x}_1 \vee a_4\bar{x}_2$

$Y11=a_5$

3. Computer Simulations

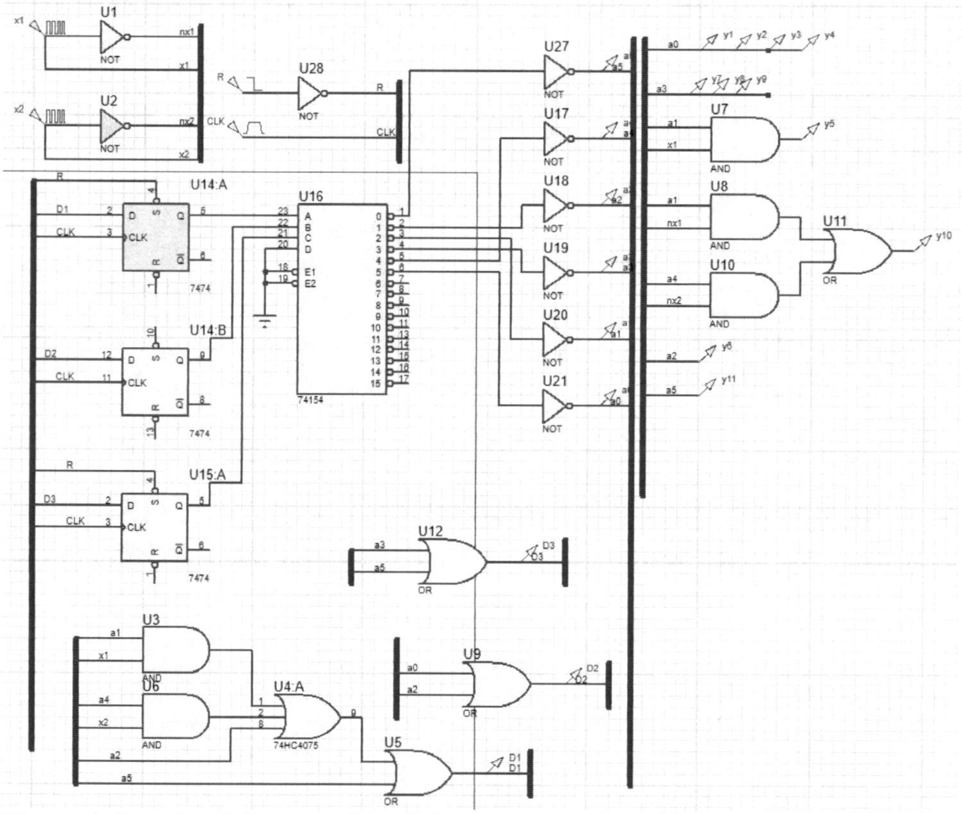

Figure 4. Functional diagram of the finite-state control machine.

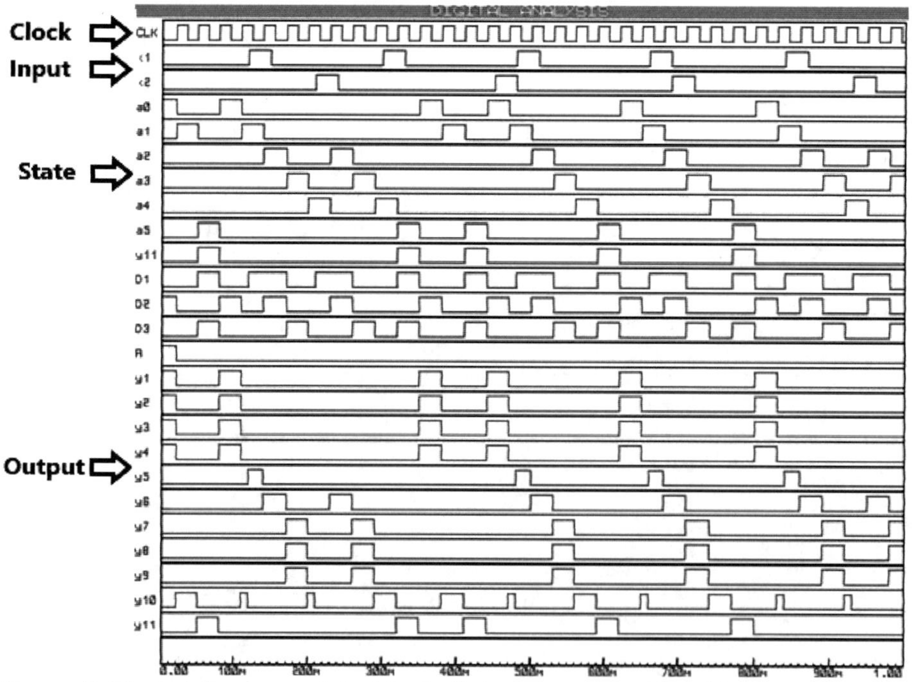

Figure 5. Timing diagrams of the Mealy finite-state control machine.

The functional diagram of the finite-state control machine is presented in Fig. 4. As the elementary base a family of 7400-series integrated circuits is used. Time diagrams are presented in Fig. 5. They show which signals are supplied to the control machine and which are sent to the operating machine for execution. The functional diagram of the datapath is quite complex and big, beyond the scope of the article. In the course of "Microprocessor engineering" students have the opportunity to implement both finite-state datapath and control machines by the means of the languages of description VHDL and System Verilog. Therefore, the student should formulate a task to synthesize according to his version a complex scheme that includes a control and operating machine, check its functionality in the CAD-environment Proteus Design Suite and then translate the synthesized complex scheme in both VHDL and System Verilog programming languages. It allows the student to program the FPGA controller with a description of its circuit. As an example, in the work the variant of the Mealy finite-state control machine with six states and logic of transitions between the states, described by the graph in fig. 3. The beginning of the VHDL-code of this machine is shown in Fig.6.

4. Hardware Implementation

Formalized hardware description language is fundamentally different from the classical programming language in that operators are not executed sequentially.At the same time it describes the logic of all signals of the electronic circuit, and consequently the logic of the device as a whole.

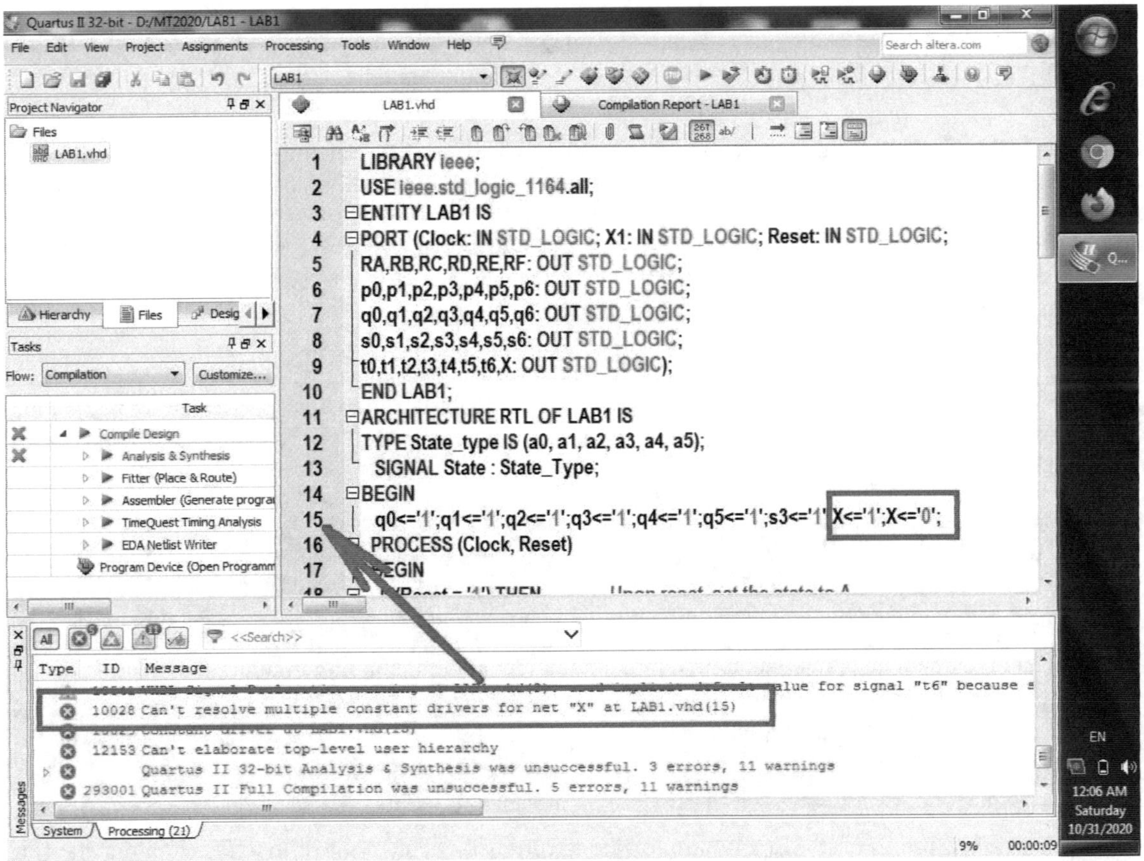

Figure 6. VHDL code Compiling Error of in Altera Quartus II CAD environment v13.0 SP1.

For example, if in the classical C-programming language two operators X = 1 are executed consecutively; X = 0; then the second operator ignores the result of the first statement and the variable X gets a final value of 0. Conversely, if the operators swap, the final value of the variable X will be 1. But if these two operators write in the hardware description language, where the first will set the signal

X units, and the second operator - logical zero, then for the CAD-environment the sequence of operators does not matter and in both cases it will return the compilation error "CAN'T RESOLVE MULTIPLE CONSTANT FOR 'X'" (Fig. 6). Only one of these statements must be transmitted for successful compilation.

```
. . . . . . . . . . . . . . . . . . . .
PROCESS (Clock, Reset)
 BEGIN
  If (Reset = '1') THEN
      State <= a0;
  ELSIF falling_edge(Clock) THEN
      CASE State IS
              WHEN a0 =>
               State <= a1;
              WHEN a1 =>
                 IF X1='1' THEN
                      State <= a2; t2<='1'; t5<='1';  t1<='0'; t4<='0';
                      p0<='1';p1<='0';p2<='0';p3<='1';p4<='1';p5<='1';p6<='1';
                      ELSE
                      State <= a5;  t1<='1'; t4<='1'; t2<='0'; t5<='0';
                      p0<='0';p1<='0';p2<='0';p3<='0';p4<='0';p5<='0';p6<='1';
              END IF;
              . . . . . . . . . . . .
      END CASE;
  END IF;
 END PROCESS;
. . . . . . . . . . . . . . . . . . . .
```

In particular, the new value type a0-a5 is declared in line 12, and the internal signal State is declared in line 13 , which can get only one of the six declared values. Signals p0-p6, q0-q6, s0-s6, t0-t0 are used for visual output of information on the first, second, third and fourth seven-segment indicator, respectively, RA-RF signals - to indicate the status of the machine using green LEDs (Fig.7). From state a1 the finite-state machine has two conditional transitions to state a2 at the value of the input signal X1 = 1 with the generation of the output signal Y5, as well as to state a5 with the generation of the output signal Y10, as shown in the transition graph (Fig.3).

From the preceding fragment of the VHDL code it is indicated that the transition from state a0 to state a1 occurs unconditionally, only on the falling edge (falling_edge) of the Clock signal and at zero value of the Reset signal. On the contrary, from state a1 there is a transition to state a2 at X1 = 1 (Fig. 7a) or to state a5 at X1 = 0 (Fig. 7b).

Figure 7. Demonstration of the functioning of the Mealy finite-state control machine implemented in the FPGA-controller Altera Cyclone II. The transition is from state a1 to state a2 (Fig. a) and from state a1 to state a5 (b).

To simplify the VHDL code, which must be written, compiled and run on the FPGA-controller during student's assignment, the signal Clock transition between states is generated manually by one of the four buttons of the controller (Fig.3). Modern FPGA-controllers also have built-in clock generators signals of almost arbitrary frequency, compatible in order of magnitude with the clock frequency of modern processors (109 Hz) and as a result of connecting the Clock signal to such a clock generator can synthesize a real functional circuit with very high speed digital finite-state machines that respond to input for several nanoseconds signals, performs a cycle of transitions between states and generates output signals.

This means that the control and operating circuit synthesized by the student and which runs on an FPGA controller is a universal blank for the development of new computing devices, including microprocessors with new specific instructions. This is the deep applied content of the problem of synthesis of such circuits, which include digital Moore or Mealy finite-state machines. Such specific computing devices are necessary to control complex devices not manufactured in series but in single copies, such as space satellites or devices operating in conditions of huge pressure (sea depths), high temperatures (volcanoes), ionizing radiation (reactors).

Even if the student or teacher does not have an FPGA controller, the Altera Quartus II integrated environment includes a built-in powerful CAD simulator ModelSIM, which analyzes the operation of the circuit, successfully compiled in the form of VHDL or System Verilog program, can simulate all possible modes of the electronic circuit. Similar to the Proteus Design Suite simulator, it also checks the correct operation and signal generation of the circuit.

Conclusions

An electronic circuit development is a complex technological procedure that contains a sequence of certain steps, including simulation of the circuit. As example of such procedure the article describes the designing steps of finite-state machine that performs the algebraic expression calculation.

The designed circuit was modeled using the Proteus Design Suite program, as well as compiled in the VHDL hardware description language and run on an Altera Cyclone II FPGA controller.

Developed remote laboratory work allows students to use modern CAD packages to design logical electronic circuits and perform computational experiments.

Created and proposed assignments for laboratory work in various disciplines, provide students with the opportunity of consistently expanding the functionality of the designed devices.

Laboratory works help students to master experience in the development and debugging of electronic circuits with a given functionality.

References

[1] Motahhir S Chalh A El Ghzizal A Sebti E G and Derouich A 2017 Modeling of Photovoltaic Panel by using Proteus *Journal of Engineering Science and Technology Review* **10(2)** 8-13

[2] Chaikivskyi T Sus B Bauzha O and Zagorodnyuk S 2020 Multicomponent analyzer of volatile compounds characterization based on artificial neural networks *CMIS-2020* **2608** 819-31

[3] Shamonia V H Semenikhina O V Proshkin V V Lebid O V Kharchenko S Ya and Lytvyn O S 2019 Using the Proteus virtual environment to train future IT professionals EducDim **53** 181-98

[4] Klimenko X A 2020 The development of digital device for current, voltage, power measuring and simulation results in proteus environment *J. Phys.: Conf. Ser.* **1515** 052050

[5] Mukherjee A Ray S and Das A 2014 Development of Microcontroller Based Speed Control Scheme of BLDC Motor Using Proteus VSM Software *IJEEE* 1-7

[6] Chaikivskyi T Sus B Tmienova N Bauzha O S and Zagorodnyuk S P 2019 3D simulation of virtual laboratory on electron microscopy *CEUR Workshop Proceedings* **2533** 282-91

[7] Tatar G Bayar S and Alkan M 2019 FPGA Based Step Motor Control For Solar Panels *2019 IEEE 13th International Conference on Application of Information and Communication Technologies (AICT)* 1-5

[8] Tang Z and Chen X 2021 VHDL Design of Motor Control Module on FPGA *Application of Intelligent Systems in Multi-modal Information Analytics* **1234** 237-45

[9] Otmani R and Benmoussa N 2020 Micro Electro Mechanical Systems Modeling by VHDL-AMS: Application to a Piezoresistive Pressure Sensor *Mechatronics 4.0* 31–41

[10] Bauzha O Sus B Zagorodnyuk S and Stuchynska N 2019 Electrocardiogram Measurement Complex Based on Microcontrollers and Wireless Networks *IEEE International Scientific-*

Practical Conference Problems of Infocommunications, Science and Technology (PIC S&T) 345-9
[11] Ayeh E Agbedanu K Morita Y Adamo O and Guturu P 2008 FPGA Implementation of an 8-bit Simple Processor *2008 IEEE Region 5 Conference* 2008 1-5
[12] Dudhane T M 2020 Architectural Enhancement of Processor with 8 Bit Multiplier and 16 Bit Co-operative ALU using VHDL *IJETER* **8** 2977–83
[13] Kung Y-S Risfendra 2016 ModelSim/Simulink co-simulation of a sensorless control for PMSM drives based on I-F startup and EKF *International Conference on Applied System Innovation (ICASI)* 1–4
[14] Shen L Zhou G and Zhai J Li H 2019 The Waveform Generator and Monitor Based on FPGA&DSP for the Pulsed Power Supply FSDM
[15] Kumar A Kumar P Modi S and Nath V 2021 Study and Implementation of Ladder Logic Conversion to VHDL for Field Programmable Gate Array (FPGA)-Based Programmable Logic Controllers (PLC) *Proceedings of the Fourth International Conference on Microelectronics, Computing and Communication Systems* **673** 31–42
[16] Garapati P and Musala S 2020 Moore and Mealy Negative Edge detector A VHDL Example for Finite State Machine *2020 International Conference on Communication and Signal Processing (ICCSP)* 1159–61

The Golden Ratio Regularities of Decayed Oscillations

P Kosobutskyy[1]*, N Jaworski[1], I Farmaha[1], U Marikutsa[1] and M Iwaniec[2]

[1]Lviv Polytechnic National University, CAD Systems Department,
79013, S. Bandera St. 12, Lviv, UKRAINE
[2]AGH University of Science and Technology, Process Control Department,
30059, al. Mickiewicza 30, Kraków, POLAND

*Email: petro.s.kosobutskyi@lpnu.ua

Abstract. This paper develops the method for linear oscillator decayed oscillations amplitude division into two unequal parts by so called golden proportion. It is shown that the relaxation time of the area under the envelope of the oscillations amplitudes and the amplitudes themselves coincide, which indicates that the decay time at the golden ratio plays the same role as the relaxation time. This simplifies the algorithm for optimization of oscillatory systems by objective function.

1. Introduction

Declared in the title task is known. In one or another way the golden ratio regularities of oscillating systems were analysed in [1-8]. This paper considers an application of the golden ratio method for analysis of regularities in changes of linear oscillator oscillations amplitude and area under their envelope. Developed model simplifies the algorithm for optimization of oscillatory systems by objective function which is the topic of interest in oscillation system computer-aided design and industrial usage, particularly in atomic-force microscopy [9-12].

2. Main results and their discussions

It is known that if the speed of linear oscillator movements is proportional to the resistance force then the oscillations amplitude envelope should decay by exponential law:

$$a_0(y) = a_0 \exp(-y), \quad y = \frac{t}{\tau}, \tag{1}$$

where τ is relaxation time. The area under envelope (1) in the positive half-plane of Cartesian coordinates is equal to:

$$S(y) = T\int_0^{\alpha} a_0 \exp(-y)dy = -\tau a_0 \exp(-y)\Big|_0^{\alpha}. \tag{2}$$

The integral value becomes to look like:

$$S_{[0,\infty)} = -\tau a_0 \exp\left(-\frac{t}{\tau}\right)\Bigg|_0^{\infty} = \tau a_0. \tag{3}$$

Let us apply the golden ratio of $S_{[0,\infty)}$ into two unequal parts $S_{[0,\tau_\phi]}$ and $S_{[0,\infty)} - S_{[0,\tau_\phi]}$ by a point τ_ϕ on the abscissa coordinate:

$$\frac{S_{[0,\tau_\phi]}}{S_{[0,\infty)}} = \frac{S_{[0,\infty)} - S_{[0,\tau_\phi]}}{S_{[0,\tau_\phi]}} \Rightarrow S_{[0,\tau_\phi]}{}^2 + S_{[0,\tau_\phi]}S_{[0,\infty)} - S_{[0,\infty)}{}^2 = 0 \Rightarrow$$
$$\Rightarrow S_{[0,\tau_\phi]}{}^2 + S_{[0,\tau_\phi]}\tau a_0 - (\tau a_0)^2 = 0 \quad S_{[0,\tau_\phi]} = \tau a_0 \begin{cases} +0.168 = \phi, \\ -1.618 = -\Phi, \end{cases} \tag{4}$$

where the numbers ϕ, Φ are known quantitative characteristics of the golden ratio [1]. Since the physical content has an area positive value, then the solution of equation:

$$S_{[0,\tau_\phi]} = \tau a_0 \cdot \phi = \tau a_0 \left(1 - \exp\left(-\frac{\tau_\phi}{\tau}\right)\right), \tag{5}$$

brings the value of τ_ϕ:

$$\exp\left(-\frac{\tau_\phi}{\tau}\right) = 1 - \phi \quad \Rightarrow \quad \tau_\phi = 0.962 \cdot \tau . \tag{6}$$

It means that after time τ_ϕ, which is only 5% less than amplitude relaxation time τ, area (2) under envelope (1) should be equal to 61.8% of total $S_{[0,\infty)} = \tau a_0$.

According to so called golden properties of roots of summarized quadratic equation:

$$x^2 + px - q = 0, \tag{7}$$

one can select the direction on the Cartesian coordinates phase plane $p0q$:

$$|p| = |q| = k, \tag{8}$$

along of which the model of the golden ratio for any division parameter L is [13-15]:

$$\frac{x}{L} = k\frac{L - x}{x}. \tag{9}$$

The phase directions (8) are parallel to so-called root lines which are shown in fig. 1.

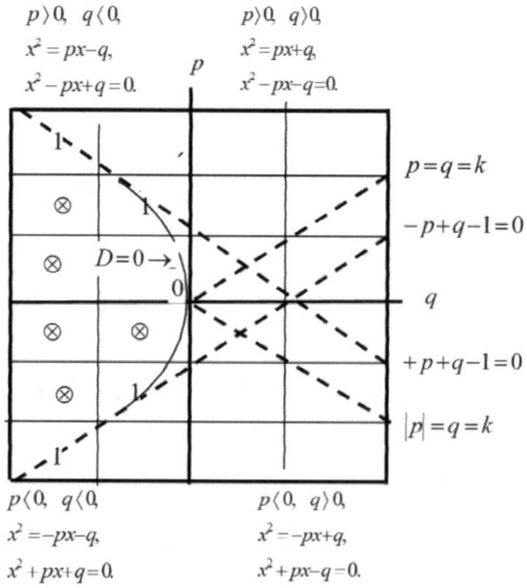

Figure 1. The first order root lines. Parabola $D = 0$ defines the area with complex roots.

Root lines are directions $K_{\pm}$ along of which the discriminant $D = p^2/4 - q$ is equal to exact square. They can be expressed as:

$$K_{\pm} = -\frac{p}{2} \pm \sqrt{D} \Rightarrow p = 0 : \begin{cases} K = \pm 1, & \left| q|_{p=0} \right| = 1, \\ K = \pm 2, & \left| q|_{p=0} \right| = 4, \\ K = \pm 3, & \left| q|_{p=0} \right| = 9, \\ \ldots \end{cases} \Rightarrow \begin{cases} +1 \cdot p + q + 1^2 = 0, \\ +2 \cdot p + q + 2^2 = 0 \\ +3 \cdot p + q + 3^2 = 0 \\ \ldots \end{cases} \Rightarrow \begin{cases} q|_{p=0, K=+1} = -1, \\ q|_{p=0, K=+2} = -4, \\ q|_{p=0, K=+3} = -9. \\ \ldots \end{cases} \quad (10)$$

Fig. 1 shows the root lines of the first order. The roots $x(p,q)_{\pm} = x_{\pm}$ along those root lines are equal to integers a_K. The sequence $\{a_K\}$ of those integers can be found by the first order recurrent expression $a_{K+1} = a_K + 1$. The coordinate system in fig. 1 is build by values of coefficients p, q and corresponding plane Opq is divided into four quadrants: $p \rangle 0, q \rangle 0$, $p \rangle 0, q \langle 0$, $p \langle 0, q \langle 0$ i $p \langle 0, q \rangle 0$. The start of reference is upper right quadrant with coordinates:

$$\begin{cases} p \rangle 0, \ q \rangle 0, \\ x^2 = px + q. \end{cases} \quad (11)$$

The second quadrant has $p \rangle 0, q \langle 0$, the third has $p \langle 0, q \langle 0$ and the fourth has $p \langle 0, q \rangle 0$.

The solutions of (9), that corresponds to fourth quadrant, shows that the bigger part of area $S_{[0,\tau_\phi(k)]}$ equals to:

$$S_{[0,\tau_\phi(k)]} = \frac{1}{2} \tau a_0 \left(-k + \sqrt{k^2 + 4k} \right), \quad (12)$$

as the function of phase parameter k. According to equation:

$$S_{[0,\tau_\phi(k)]} = \frac{1}{2} \tau a_0 \left(-k + \sqrt{k^2 + 4k} \right) = \tau a_0 \left(1 - \exp\left(-\frac{\tau_\phi(k)}{\tau} \right) \right), \quad (13)$$

the value of an arbitrary division coordinate $\tau_\phi(k)$ in model (8) can be described by logarithmic function:

$$\exp\left(-\frac{\tau_\phi(k)}{\tau} \right) = 1 + \frac{k - \sqrt{k^2 + 4k}}{2} \Rightarrow \tau_\phi(k) = \tau \ln\left(\frac{2}{2 + k - \sqrt{k^2 + 4k}} \right). \quad (14)$$

E.g., table 1 shows the values of area (13) for time intervals (14), which are expressed through discrete integer and so called golden values k, which generally can be of any non-integer or irrational ones.

Table 1. The values of area $S_{[0,\tau_\phi(k)]}$ for time intervals $\tau_\phi(k)$ through golden values k.

k	1	2	3	4	5
$\tau_\phi(k) \cdot \tau^{-1}$	0.962	1.317	1.567	1.763	1.925
$S_{[0,\tau_\phi(k)]} \cdot S_{[0,\infty)}^{-1}$	0.618	0.723	0.791	0.828	0.854

Let us find the formula for the golden division of oscillations amplitude maximum $a(y) = a_0 exp(-y)$:

$$\frac{a_{\tau_\varphi}}{a_0} = \frac{a_0 - a_{\tau_\varphi}}{a_{\tau_\varphi}} \Rightarrow \left(\frac{a_{\tau_\varphi}}{a_0}\right)^2 + \frac{a_\tau}{a_0} - 1 = 0 \Rightarrow a_{\tau_\varphi} = a_0\phi . \quad (15)$$

Time τ_φ during of which relative amplitude a_{τ_φ} / a_0 decreases to value ϕ can be expressed as:

$$a_0\phi = a_0 exp\left(-\frac{\tau_\varphi}{\tau}\right) \Rightarrow \tau_\varphi = 0.481 \cdot \tau . \quad (16)$$

After time τ_φ the oscillations amplitude is approximately 48% of its starting value. The change of relative area after this time obtains magnitude of $\phi^2 = 0.382$. The relation of two time intervals τ_ϕ and τ_φ equals:

$$\frac{\tau_\phi}{\tau_\varphi} = \frac{0.962}{0.481} = 2 , \quad (17)$$

i.e. τ_φ is two times less then τ_ϕ. The relative change of amplitude also obtains magnitude of $\phi^2 = 0.382$ after time τ_ϕ. It is because the plane (5) was calculated only in the half-plane for one equilibrium deviation direction of oscillator. This value should be multiplied by two. That is why characteristics of τ_φ and are equal. The relative change of amplitude for other relative values of time interval are described in table 2.

Table 2. The relative change of amplitude $a(y)$ for other relative values of time interval.

$a(t/\tau)$	ϕ	0.607	0.5	ϕ^2
$\ln\left(a(t/\tau)^{-1}\right)$	0.481	0.5	0.693	0.963

For the points with an arbitrary value of k (table 3) the following relation is true:

$$S(y) + [a_0(y) \cdot dy] = a_0 \overline{\tau_{[0,\infty)}} = a_0\tau , \quad (18)$$

where amplitude values $a_0(y)$ are multiplied by a unit value $dy = 1$. The coordinates of division are chosen as the middle values of time intervals during of which amplitudes were changed as:

$$\alpha = \frac{\upsilon}{\tau} \Rightarrow \begin{cases} \upsilon = \overline{\tau_{[0,\tau_\phi]}}: & \alpha = 0.405: \dfrac{a_{\tau_\phi}}{a_0} = \dfrac{\tau_\phi/\tau \cdot e^{-\tau_\phi/\tau}}{1 - e^{-\tau_\phi/\tau}}, \\ \upsilon = \overline{\tau_{[0,\tau]}}: & \alpha = 0.418: \dfrac{a_\tau}{a_0} = 1 - \dfrac{1 - 2e^{-1}}{1 - e^{-1}}. \end{cases} \quad (19)$$

Table 3. The values of area $S(y)$ and amplitude α for division coordinates as the middle values of time intervals

α	$\overline{\tau_{[0,\tau_\phi]}} = 0.405 \cdot \tau$	$\overline{\tau_{[0,\tau]}} = 0.418 \cdot \tau$	$\tau_\phi = 0.962 \cdot \tau$
$S(\alpha) \cdot S_0 = 1 - e^{-\alpha}$	0.333	0.342	0.618
$a(\alpha) \cdot a_0 = e^{-\alpha}$	0.667	0.658	0.382

Let us consider a division of linear oscillator decayed oscillations amplitude in area of arbitrary half-period $a\left(\tau_{\phi,n}=nT/2\pm\delta_{\phi,n}\right)=a_0e^{-nT/2}\sin\left(n\pi\pm 2\pi\delta_{\phi,n}/T\right)$:

$$\frac{a(\tau_{\phi,n})}{a_{0n}}=\frac{a_{0n}-a(\tau_{\phi,n})}{a(\tau_{\phi,n})}\Rightarrow\frac{e^{-n\frac{T}{\tau}}\sin\left(2n\pi\right)-e^{-n\frac{T\pm\delta_{\phi,n}}{\tau}}\sin\left(2n\pi\pm\omega_0\delta_{\phi,n}\right)}{e^{-n\frac{T\pm\delta_{\phi,n}}{\tau}}\sin\left(2n\pi\pm\delta_{\phi,n}\right)}=$$

$$=\frac{e^{-n\frac{T\pm\delta_{\phi,n}}{\tau}}\sin\left(2n\pi\pm\omega_0\delta_{\phi,n}\right)}{e^{-n\frac{T}{\tau}}\sin\left(2n\pi\right)}\Rightarrow\frac{\sin\left(2n\pi\right)-e^{-n\frac{\delta_{\phi,n}}{\tau}}\sin\left(2n\pi\pm\omega_0\delta_{\phi,n}\right)}{\sin\left(2n\pi\pm\delta_{\phi,n}\right)}=, \tag{20}$$

$$=\frac{e^{-n\frac{\delta_{\phi,n}}{\tau}}\sin\left(2n\pi\pm\omega_0\delta_{\phi,n}\right)}{\sin\left(2n\pi\right)}\Rightarrow\frac{e^{-n\frac{\delta_{\phi,n}}{\tau}}\sin\left(2n\pi\pm\omega_0\delta_{\phi,n}\right)}{\sin\left(2n\pi\right)}=\phi,$$

where time $\delta_{\phi,n}$ is counting from half-period amplitude maximum n. The value can be obtained as:

$$\sin\left(\omega_0\delta_{\phi,n}\right)=\phi\Rightarrow\delta_{\phi,n}=\frac{1}{\omega_0}\arcsin\phi\Rightarrow\frac{\delta_{\phi,n}}{T}=\frac{\arcsin\phi}{2\pi}=0.106\,. \tag{21}$$

The last value is near 33% of relative time $\tau_\varphi/\tau=0.481$ of amplitude golden relaxation in area of the first maximum after oscillations beginning. The corresponding initial conditions for the oscillations appearance are that at the starting moment oscillator is deviated and stays in equilibrium.

Conclusions

This paper develops the method for linear oscillator decayed oscillations amplitude division into two unequal parts by so called golden proportion. It is shown that the relaxation time of the area under the envelope of the oscillations amplitudes and the amplitudes themselves coincide, which indicates that the decay time at the golden ratio plays the same role as the relaxation time. This simplifies the algorithm for optimization of oscillatory systems by objective function.

References

[1] Gazale M 1999 *Gnomon. From Pharaohs to Fractals* (Princeton New Jersey: Princeton University Press)
[2] Rosson B 2010 Characteristic equations with solutions that contain natures golden ratio *WIT Transactions on Ecology and the Environment* **138** 153–63 https://doi.org/10.2495/DN100141
[3] Kosobutskyy P 2019 Modelling of electrodynamic systems by the method of binary separation of additive parameter in golden proportion *Journal of Electronic Research and Application* **3(3)** 8-12 https://doi.org/10.26689/jera.v3i3.807
[4] Pashev O and Nalci S 2012 Golden quantum oscillator and Binnet-Fibonacci calculs *Journal of Physics A: Mathematical and Theoretical* **45(1)** 015303-15 https://doi.org/10.1088/1751-8113/45/1/015303
[5] Moorman C and Goff J 2007 Golden ratio in a coupled-oscillator problem *European Journal of Physics* **28** 897-902 https://doi.org/10.1088/0143-0807/28/5/013
[6] Daqiu Y Dongfeng X and Henryk R 2006 Golden ratio and bond-valence parameters of hydrogen bonds of hydrated borates *Journal of Molecular Structure* **783** 210-4 https://doi.org/10.1016/j.molstruc.2005.08.022
[7] Kharitonov A 2007 Structural exposition of composite systems *Applied Physics (Russia)* **1** 5-9
[8] Forbes K Williams M and Andrews D 2016 Quantum theory for the nanoscale propagation of the light through stacked thin films film layers, *SPIE Photonics Europe* **9884** 988434-6 https://doi.org/10.13140/RG.2.1.3250.4724

[9] Kosobutskyy P Jaworski N Farmaha I and Kuzmynykh M 2019 Optimization of probe parameters of atomic force microscope cantilever *2019 IEEE XVth International Conference on the Perspective Technologies and Methods in MEMS Design (MEMSTECH)* 127-130 https://doi.org/10.1109/MEMSTECH.2019.8817389

[10] Kosobutskyy P Jaworski N and Lobur M 2017 Analysis of cauchy criterion similarities in miniaturization of elastico-dynamic cantilever systems *2017 XIIIth International Conference on Perspective Technologies and Methods in MEMS Design (MEMSTECH)* 6-8 https://doi.org/10.1109/MEMSTECH.2017.7937519

[11] Lobur M Kosobutskyy P Holovatskyy R Jaworski N and Farmaga I 2017 The atomic-force nicroscope — project of the CAD department of Lviv polytechnic National university *2017 14th International Conference The Experience of Designing and Application of CAD Systems in Microelectronics (CADSM)* 402-4 https://doi.org/10.1109/CADSM.2017.7916161

[12] Maisakovskyi R Kolesnyk K Panchak R and Golovatskyy R 2016 Automated simulation of scanning tunneling microscope functional parameters XII International Conference on Perspective Technologies and Methods in MEMS Design (MEMSTECH) 107-9 https://doiorg/10.1109/MEMSTECH.2016.7507530

[13] Kosobutskyy P 2020 Phidias numbers as a basis for Fibonacci analogues *Notes on Number Theory and Discrete Mathematics* **26(1)** 172-8 https://doi.org/10.7546/nntdm.2020.26.1.172-178

[14] Kosobutskyy P Karkulovska M and Morgulis A 2018 Mathematical methods for CAD: the method of proportional division of the whole into two unequal parts *Bulletin of the Lviv Polytechnic National University: Computer Design Systems. Theory and practice* **908** 75

[15] Kosobutskyy P 2018 On the possibility of constructing a set of numbers with golden section properties *International Conference Algebra and Analysis with Application* 1-4

IOP Conf. Series: Materials Science and Engineering 1016 (2021) 012008 doi:10.1088/1757-899X/1016/1/012008

Mathematical Model of Rheological Behavior of Wood Plate Taking Into Account the Zone of Evaporation of Moisture

Ya Sokolovskyy*, I Boretska, I Kroshnyy, S Yatsyshyn and O Storozhuk

IT Department, Ukrainian National Forestry University, Lviv, UKRAINE

*Email: sokolowskyyyar@yahoo.com

Abstract. The subject of this paper is a mathematical model of heat-transfer in the process which takes place in capillary-porous materials during drying. The moving boundaries of phase transitions for media are also considering. The analytical-numerical method was developed for mathematical model. The difference schemes were constructed with consideration for the boundaries of the evaporation zone of heat exchange with the environment in the presence of a phase transition in the conditions. The numerical realization and software were developed for solving the Stefan problem. The fractal structures of the medium were taking into account for synthesization of mathematical models of viscoelastic deformation. Functional dependences of the temperature field and the motion of the phase transition boundary in media, depending on the value of fractional parameters, were received for modelling different technological processes of drying capillary-porous materials.

1. Introduction

The intensification of drying schedules for colloidal capillary-porous materials leads to further development of mathematical modeling of the processes of heat-and-mass transfer, phase transformation, and deformation, which would adequately describe the laws of moisture removal in the materials being dried. The presence of a moving boundary of phase transformations at the interface between phases with different thermophysical and mechanical characteristics considerably complicates mathematical models of the processes of deformation-relaxation and heat-and-mass exchange during the drying of capillary-porous materials. The simulation of heat-and-mass transfer with phase transitions in the drying process is reduced to solving Stefan's problems, which are the most complicated even for minor changes in the density of the material in the evaporation zone. Therefore, the mathematical modeling of such processes by generalizing the Stefan problem for media requires the development of new non-traditional methods and modeling tools.

The general approach to modeling the interconnected heat-and-mass transfer and the stress-strain state is based on the thermodynamics of irreversible processes. In research works [1,4] this approach is applied to the study of the drying process of capillary-porous materials. In the framework of the theory of small viscoelastic-plastic deformations, a one-dimensional problem of self-stresses in a thin free plate during symmetric drying on both sides was solved. In the work [8], the equations of heat-and-mass transfer in deformed elastic capillary-porous media were proposed, taking into account the dependence of surface properties on temperature. However, they do not describe the presence of irreversible deformations due to the shrinkage process.

In all these studies, the type of rheological behavior of the material is assumed to be independent of moisture. However, in many cases the rheological characteristics of the medium in a dry and a saturated state are significantly different. In this regard, it is possible to distinguish the series of researchers' works [3].

The development of mathematical models of viscoelastic capillary-porous materials in the drying process is based on hereditary creep models [4]. The equation of the balance of energy of dissipation on the surface of the phase transition makes it possible to formulate the conditions for the deepening

IOP Conf. Series: Materials Science and Engineering 1016 (2021) 012008 doi:10.1088/1757-899X/1016/1/012008

zone of evaporation, taking into account the preservation of irreversible deformations during phase transitions. However, such models describe the rheological behavior of various media for a continuous history of deformation. If, in the process of deformation of capillary-porous media, phase transitions are implicit and the process of deformation at their boundary is characterized by jump discontinuities, it is necessary to take into account the influence of the prehistory of loading before the phase transition on the further development of the stress-strain state of the medium.

2. A mathematical model of the rheological behavior of a wood plate taking into account the moisture evaporation zone

The stress-strain state arising in the process of drying wood is the main factor in slowing down the intensification of the technological process. The complexity of calculating the components of stress and deformation is due, primarily, to the behavior of the material, the anisotropy of its structure, the capillary-porous structure of the medium. The process of drying wood, especially during a period of decreasing drying speed, is characterized by the presence of at least two zones of moisture evaporation with different technological characteristics.

The paper deals with the mathematical model of the rheological behavior of a wood plate in the process of convection drying, taking into account the moisture evaporation zone. In the moist area of the wood plate, the material is characterized by viscoelastic properties. To study the stress-strain state in the moist zone, the Kelvin-Voigt model is used. The dried zone is described by elastic behavior. One of the causes of stresses in the materials being dried is free drying conditioned by capillary processes of free (intercapillary) moisture removal and the occurrence of uneven drying over the thickness of the material.

The total deformation of wood under conditions of convection drying can be represented in the for

$$\varepsilon = \varepsilon_{\ddot{I}} + \varepsilon_{\hat{A}} + \varepsilon_{W}, \tag{1}$$

where $\varepsilon_{\ddot{I}}$, ε_{B}, ε_{W} are, respectively, elastic, visco-elastic, and shrinkage deformations.

To study elastic deformations, Hooke's law is used, taking into account shrinkage deformations which are determined by a linear dependence on changes in moisture content. For the moist zone of the wood plate, the Kelvin-Voigt rheological model is described by the relation [5]

$$GK\left(1+\tau_1\frac{\partial}{\partial\tau}\right)\left(1+\tau_0\frac{\partial}{\partial\tau}\right)\left(\varepsilon_{ij}-\frac{\varepsilon_w}{3}\delta_{ij}\right)=3K\left(1+\tau_0\frac{\partial}{\partial\tau}\right)\sigma_{ij}- \tag{2}$$

$$-\left(3K\left(1+\tau_0\frac{\partial}{\partial t}\right)-\frac{2}{3}G\left(1+\tau_1\frac{\partial}{\partial\tau}\right)\sigma_{ii}\delta_{ij}\right),$$

where ε_{ij}, σ_{ij} are components of the strain and stress tensor; G, K are module of elasticity and volume expansion, τ_0, τ_1 are relaxation time, δ_{ij} are Kronecker's symbols.

In the case of a one-dimensional moisture field during the drying of a wood plate with a thickness of $2l$ taking into account the dependencies $\sigma_z = 0$,, $\sigma_x = \sigma_y = \sigma_z$, $\varepsilon_x = \varepsilon_y = \varepsilon_z$ from (2) we get the mathematical model in the form:

$$\tau_0\tau_1\frac{\partial^2\varepsilon_B(\tau)}{\partial\tau}+(\tau_0+\tau_1)\frac{\partial\varepsilon_B(\tau)}{\partial\tau}+\varepsilon_B(\tau)=\frac{4G\tau_1+3K\tau_0}{18GK}\frac{\partial\sigma_B}{\partial\tau}+\frac{4G+3K}{4G\tau_1+3K\tau_0}\sigma_B. \tag{3}$$

The boundary conditions are characterized by the absence of stresses and deformations in the wood plate at the beginning of the drying process $(\tau)=0$ and take the form $\sigma = 0$, $\varepsilon = 0$.

In (3) it was taken into account that in the moist zone, shrinkage of wood is insignificant, therefore $\varepsilon_W = 0$.

The stress-strain state of the wood plate in the dried zone, taking into account the shrinkage, according to the accepted assumptions, is described by law [5, 6]

$$\sigma_{ij} = \frac{E}{1+\nu}\left(\left(\varepsilon_{ij} - \varepsilon_W\right) + \frac{\nu}{1-2\nu}\delta_{ij}\left(\varepsilon_{kk} - \varepsilon_w\right)\right) \tag{4}$$

Since the one-dimensional problem of the drying process is considered, the compatibility equation [5] can be represented:

$$\frac{d^2}{dt^2}\left(\sigma + \frac{E}{1+\nu}\right)\varepsilon_W = 0; \quad \frac{\partial^2 \varepsilon}{dz^2} = 0; \tag{5}$$

Taking into account (5) with (4), we obtain

$$\sigma(\tau) = \frac{E}{1-\nu}\left(\varepsilon(\tau) - \varepsilon_w\right). \tag{6}$$

According to [6, 7], the process of transition from the moist zone to the dried zone during the process of wood drying is characterized by almost constant values of shrinkage, then we assume that the value of shrinkage deformation ε_W is equal to the total deformation (1) at the time of transition τ^*, between elastic zones, that is $\varepsilon_W = \varepsilon^0(\tau) = \varepsilon(\tau^*)$, $\tau^* = \xi^{-1}(x)$, where $\xi(x)$ is the coordinate of the phase boundary surface which is parallel to the plane of the wood plate. Thus, the shrinkage deformation, in addition to the processes of deformation of the wood plate during the drying process, also takes into account the size of the phase transition $\xi(x)$ depending on the technological modes of the drying process, in particular non-isothermal moisture transfer.

The boundary conditions on the surface of the wood plate, given that the moment of forces and total forces are equal to zero, take the form:

$$\int_{-e}^{e} \sigma(z)dz = 0, \quad \int_{-e}^{e} z\sigma(z,\tau)dz = 0. \tag{7}$$

Thus, integrating equation (4) with zero boundary conditions $(\tau = 0 : \varepsilon = 0, \partial\varepsilon/\partial\tau = 0)$ for the moist zone and taking into account relations (6), (7) for the dried zone, we obtain a mathematical model for determining the stress-strain state of a wood plate during convection drying taking into account the moving zone of moisture evaporation in the form $z \le z^*$:

$$\sigma(\tau) = A_0\,\varepsilon(\tau) + A_1\frac{\partial\varepsilon(\tau)}{\partial\tau} + A_2\int_0^{\tau}\varepsilon(\tau')\exp(-b_0(\tau-\tau'))d\tau', \tag{8}$$

where the following notation is introduced:

$$A_0 = \frac{E(\tau_0+\tau_1)b_1 - 3(1-\nu)\tau_1\tau_0}{b_1^2};$$

$$A_1 = \frac{E\tau_0\tau_1}{b_1}; A_2 = \frac{E}{b_1^3}\left(b_1^2 - 3(1-\nu)(\tau_1+\tau_0)\left(b_1 + 9(1-\nu)^2\tau_0\tau_1\right)\right);$$

$$b_0 = \frac{3(1-\nu)}{2(1-2\nu)\tau_1 + (1+\nu)\tau_0}; \quad b_1 = 2\tau_1(1-2\nu) + \tau_0(1+\nu).$$

For the dried zone $z^* \le z \le l$ we can write down accordingly

$$\sigma = \frac{E}{1-\nu}\left(\varepsilon + \frac{\varepsilon_W}{3}\right). \tag{9}$$

In (8), (9) by E and ν denote the elastic module and Poisson's ratios for wood which in the general case are functions of moisture content and temperature, i.e. $E = E(U,t)$, $\nu = \nu(U,t)$.

The determination of these values characterizing the rheological behavior of wood is associated with the analysis of experimental data. For this purpose, we use the experimental dependences of the creep deformation curves for wood across the fibers for different temperatures and moisture content [6, 9]. Implementation features of complex structured composite materials effective mechanical characteristics finding subsystem that is based on microlevel cellular models usage are described in papers [11, 12].

To determine the deformations ε_W during the drying of the wood plate, which are included in (8), (9) taking into account the moisture evaporation zone, we used the approach [7] and the results of temperature and moisture modeling from [13, 14] .In particular, for the case of one-dimensional mass-transfer fields, we have the following expressions for deformations in the moist and dried zones of the plate]

$$\varepsilon_W = \beta_1 \Delta U^{(1)} = \beta_1 \Delta(U_1(z,\tau) - U_0), \quad 0 < z \le z^*; \tag{10}$$

$$\varepsilon_W = \beta_2 \Delta U^{(2)} = \beta_2 \Delta(U_2(z,\tau) - U_0), \quad z^* < z \le l,$$

where β_i, $U_i\,(i = 1,2)$ are the corresponding coefficients of shrinkage and the corresponding values of moisture content, U_0 is moisture content at the beginning of the drying process. Moisture content U is defined as the ratio of the mass of moist wood to the mass of absolutely dry material.

The important point is to determine the shrinkage coefficients β_i. For wood, the decisive coefficient is β_2 for the drying zone which is associated with the removal of moisture under the action of capillary forces. Therefore, the shrinkage coefficient is determined by the movement the boundary of the evaporation zone. Note that the geometric dimensions of the moisture evaporation zone are not taken into account. In view of this, the dependence [1] is accepted for quantitative description $\varepsilon_M = \beta_0 V_0 (V + V_0)$, where β_0 is the shrinkage coefficient for low-intensive drying processes, as, for example, for a period of constant drying rate. It is assumed that [8] the speed of movement of the phase transition boundary is equal to $V = -dz^*/d\tau = j/\rho$, where j is the flow of moisture that is inversely proportional to the size of the dried zone. It can be determined by the formula [4,8]: $j = A/(1 - z^*/l)$, where $A = l/2t_1$ is some constant which depends on the thermophysical characteristics of ambient humidity: $A = [D]MP_s/RTl$. Such assumptions make it possible to make transition from time to coordinate of the evaporation boundary by the formula $\xi_* = 1 - z^*/l = (2A\tau/l)^{1/2}$, since $\xi_*(\tau) = \tau^{1/2}$, $d\xi_*(\tau)/d\tau = (2\xi_*)^{-1}$. In turn, this allows determining the deformations $\varepsilon(\xi_*)$ at the phase transition boundary from the boundary conditions (7).

To determine the deformations, the rheological equation [10] is used in the form

$$\partial^2\varepsilon/\partial\tau^2 + \alpha_1 \partial\varepsilon/\partial\tau + \alpha_2 \varepsilon = \alpha_3 \tag{11}$$

The parameter values in this equation take the form

$$\alpha_1 = 2\frac{\tau_k}{\tau_1\tau_2}(1 - \xi/l_2)(\tau_1 + \tau_2 - (1 - \xi/l_2)) \cdot (4\tau_2(1-2\nu) + 3\tau_1(1+\nu))/2(1-\nu), \tag{12}$$

$$\alpha_2 = 4\frac{\tau_k}{\tau_1}(1-\xi/l_2)^2 \cdot \left(1+\frac{l_2}{\xi}\frac{10\nu+14}{9(1-\nu)}\left(1-\frac{\xi}{l_2}+\frac{4\tau_2(1-2\nu)+3\tau_1(1+\nu)}{2\tau_k(7-5\nu)(1-\xi/l_2)\xi/l_2}\right)\right),$$

$$\alpha_3 = \frac{4}{81}\cdot\frac{4\tau_2(1-2\nu)+3\tau_1(1+\nu)}{\tau_2(1-2\nu)}\cdot\frac{\tau_k}{\tau_1}\left(1-\frac{\xi}{l_2}\right)^2 \cdot$$

$$\cdot\frac{2(11+15\nu)/(8\tau_2(1-2\nu)+6\tau_1(1+\nu)))(1-\xi/l_2)(\xi/l_2)+(\tau_2/\tau_k)}{2(1-\xi/l_2)(\xi/l_2)^2}+$$

$$+\left(1-\frac{\xi}{l_2}-\frac{l_2}{2\tau_k\theta}\right)\ln\left(1-\frac{\xi}{l_2}-\frac{l_2}{2\tau_k\theta}\right)+\frac{\tau_2/\tau_k}{2(1-\xi/l_2+l_2/2\tau_k\theta)(\xi/l_2)}.$$

3. Simulation results

The dependencies for determining the shrinkage coefficients $\beta_1=\beta_2=\beta$ take the form [8]. The value β_1 in the moist zone is constant and calculated according [8] for $U=0{,}25\div0{,}3$ and $t=20$ ^{0}C depending on the species of wood.

For the numerical experiment, pine wood with a base density $\rho_0=450$ kg/m^3 was selected for the study, using the mathematical models obtained above, the viscoelastic behavior of the wood plate during convection drying. The parameters of the drying agent were as follows: initial moisture content $u_0=0{,}4$ kg/kg; initial temperature T_0=20 ^{0}C; ambient temperature t_c = 70 ^{0}C; relative humidity $\varphi_0=60\%$; the speed of movement of the drying agent v =2 m/s.

Fig. 1 shows the graphical dependences of the stresses σ on the thickness of the wood plate depending on the different values of the moisture evaporation boundary ξ_* for different values of drying time. Fig. 2 characterizes the change in stresses during the drying of a wood plate at the evaporation boundary for various moisture content values. Fig. 3 show graphical dependences of the change of stresses in the wood plate for various values of drying time, initial moisture content, and different operating parameters of drying. Fig. 4 characterizes the change of relative deformations in the wood plate for various values of drying time (1-Fo=2,15, 2-Fo=1,15, 3-Fo=0,60, 4-Fo=0,45).

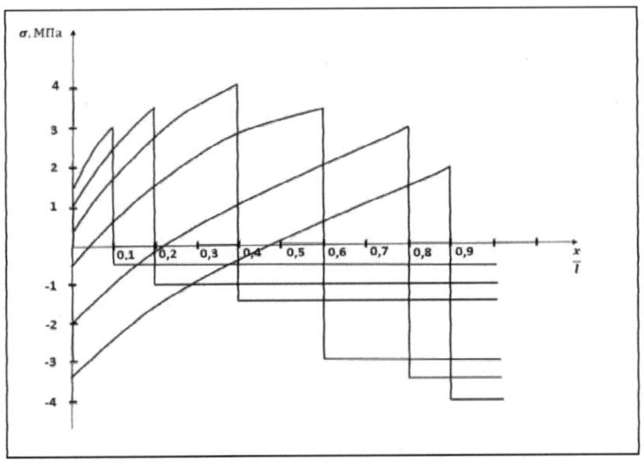

Figure 1. Change in stresses σ over the thickness of the wood plate for different values of ξ_*.

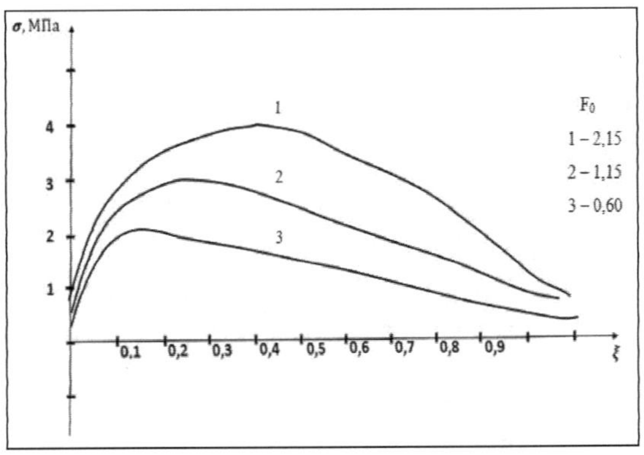

Figure 2. Change in stresses σ in the wood plate for various values of drying time.

IOP Conf. Series: Materials Science and Engineering 1016 (2021) 012008 doi:10.1088/1757-899X/1016/1/012008

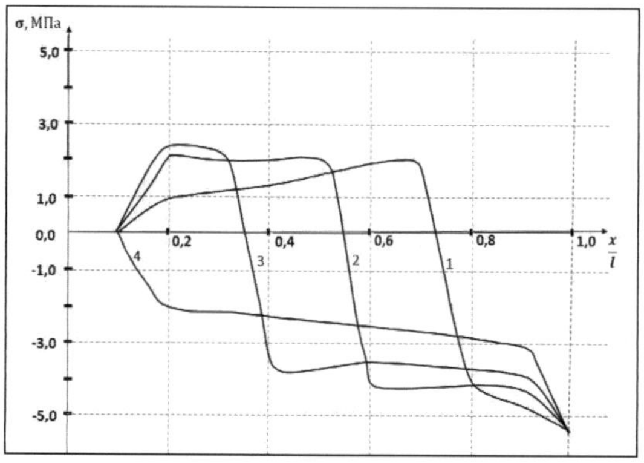

Figure 3. Change in stresses σ in the wood plate for various values of drying time.

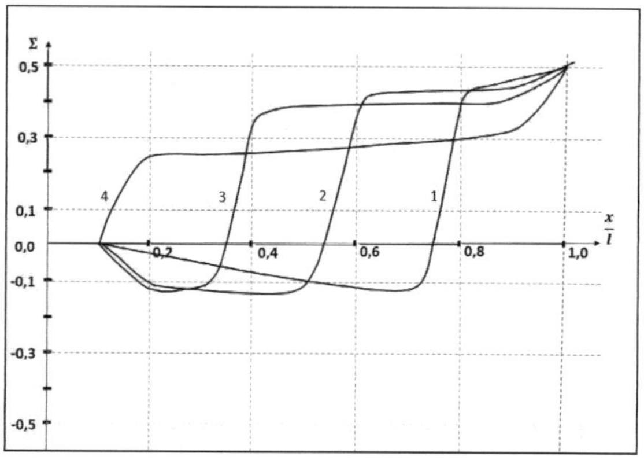

Figure 4. Change in deformation ε in the wood plate for various values of drying time.

Conclusions

A new mathematical model of viscoelastic deformation of a capillary-porous plate under conditions of changing moisture transfer with taking into account the moisture evaporation zone is constructed. A study of stresses and strains in the moist and dried areas of the wood plate was conducted. It was shown that stresses and strains for various values of the evaporation zone are characterized by time non-monotony, a shift in the values of the maximum stresses appears in the moist zone with an increase in the drying time and a jump discontinuity at the phase boundary. The patterns of the development of stress components in wood during drying are determined depending on changes in temperature, moisture content, moisture evaporation zone and technological parameters of the drying agent. The presence of compressive stresses in the dried zone of the plate for the initial stages of the drying process was detected.

References

[1] Dornyak O R and Shamaev V A 2011 Mathematical modeling of theintensive heat and mass transfer processes during convective drying of wood *Bulletin of Voronezh State Technical University* **12-1б(7)** 85-91

[2] Sychevskyi V A 2013 Stress-deformed state of wood during drying *Energy, News of higher educational institutions and CIS energy associations* **4** 67-73

[3] Siasev A V and Kostaschuk M V 2015 Modeling the process of core crystallization taking into account the mutual influence of temperature and mechanical fields *Bulletin of Dnipropetrovsk University Series: Modeling* **8(23)** 9-28

[4] Akulych P V 2010 *Calculations of drying and heat exchangers* (Minsk: Belarusian science) p 436

[5] Molotnikov V Y and Molotnikova A O 2017 *The theory of elasticity and plasticity* (St Petersburg: Lan) p 532

[6] Glukhikh V N Akopyan A L and Okhlopkova A Yu 2018 *Natural features of wood* (St Petersburg: Polytechnic University) p 391

[7] Hayvas B I 2010 Mathematical modeling of convective drying of materials taking into account mechanotherm diffusion processes *Scientific collection Physics and mathematics. modeling and information technology* **12** 9-37

[8] Minenkov V A and Shevchenko D V 1998 Modeling of stresses and strains during convective drying of a thin dispersed plate *Engineering Physics Journal* **32(2б)** 147-52

[9] Cruz-Lefevre M Remond R Aleon D and Perré P 2010 Effect of oscillating drying conditions on variations in the moisture content field inside wood boards *Wood Material Science & Engineering* **5** 84-90

[10] Sokolovskyy Ya Boretska I Yatsyshyn S and Kaspryshyn Y 2018 Mathematical modeling of deformation-relaxation processes under phase transition *CEUR Workshop Proceedings* **2300** 83-6

[11] Jaworski N Iwaniec M and Lobur M 2019 Implementation features of composite materials effective mechanical characteristics finding method based on microlevel cellular structural models *15th International Conference on the Experience of Designing and Application of CAD Systems (CADSM)* 1-5 https://doi.org/10.1109/CADSM.2019.8779273

[12] Jaworski N Andrushchak N Lobur M Iwaniec M and Sitek W 2019 Composite Material Microlevel Cellular Model Data Transfer and Verification by FEM Analysis Software Systems *IEEE XVth International Conference on the Perspective Technologies and Methods in MEMS Design (MEMSTECH)* 17-22 https://doi.org/10.1109/MEMSTECH.2019.8817376

[13] Sokolovskyy Ya Boretska I and Gayvas B 2018 Mathematical modeling of moving boundaries of phase transition in the process of drying anisotropic plate *Journal of Engineering Research and Application* **8(12)** 57-63

[14] Sokolovskyy Ya Boretska I Gayvas B and Kroshnyy I 2018 Mathematical modeling of the heat-mass-exchange in anisotropic environments taking into account the boundary of phase transition *Computer Science and Information Technologies CSIT* 147-150

IOP Conf. Series: Materials Science and Engineering 1016 (2021) 012009 doi:10.1088/1757-899X/1016/1/012009

Design Model for Determination of the Strength Fiber-Reinforced Concrete

Viktor Sylovanyuk[1], **Andriy Lisnichuk**[1]
Kostiantyn Kolesnyk[2] **and Roman Panchak**[2]

[1]Karpenko Physico-Mechanical Institute of the National Akademy of Sciences of Ukraine, 79053, Naukova str. 5, Lviv, Ukraine
[2]Lviv Polytechnic National University, CAD Systems Department, 79013, S. Bandera St. 12, Lviv, UKRAINE

*Email: kostyantyn.k.kolesnyk@lpnu.ua

Abstract. The model of fiber-reinforced concrete for determination of strength under tensile was developed. Formula for assessment of the strength of fiber-reinforced concrete, which takes into account micro-cracks and pores in the materials structure and the presence of reinforcing fibers, was established on the basis of the model. Experimental study for establishment of strength of cement reinforced with fiber of different nature and different volume content were conducted to confirm the received formula. It was shown that theoretical prediction of fiber-reinforced concrete strength correlates with experimental data.

1. Introduction

The structural material that is able to solve a set of the most complex problems in the construction industry is fiber-reinforced concrete (fiber concrete). It is a composite based on a cement matrix, reinforced with different fibers, (steel, glass, basalt, carbon, polymer and some others). Today, three types of fiber reinforced materials are chiefly widely used: fibers in the form of short pieces of thin steel wire, glass fibers and fibers based on polypropylene (Fig. 1). A broad use of reinforced concrete is due to higher strength, crack resistance, toughness, less microcracking during hardening compared to conventional concrete [1].

Considering concrete as a conglomerate of cement stone, sand and gravel, three levels of its structure are distinguished:

- microstructure – structure of cement stone;
- mesostructure – the structure of cement-sand mortar;
- macrostructure – the structure of gravel and mortar.

Each of these structures contributes to the strength of concrete, but it is largely determined by the strength of cement stone. The modelling of such materials is a topical task [2-7].

The work is devoted to modelling the processes of cracking, deformation and fracture of cement-matrix composites reinforced with various types of microfibres. Computational models are developed for prediction of tensile strength, compressive strength and fracture (behaviour) of these composites. The models are based on fracture mechanics approaches, in particular on the δ_c-model and the power criterion of the critical stress intensity factors [8].

The analytical dependences of strength and fracture toughness of the material on the mechanical properties of the phases and their volume content, the matrix porosity characteristics, etc. are obtained. The influence of the material porosity, presence of cracks in it and the volume content of the reinforcing fibers in the composites on their service characteristics are analyzed.

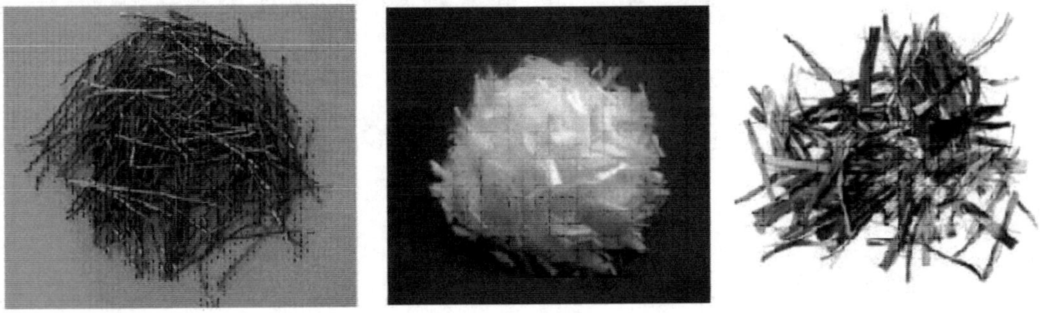

Carbon fiber Glass fiber Basalt fiber

Figure 1. Different types of reinforcing fibers.

2. Main results and their discussions

The aim of the work is to develop a mathematical model of composites based on a cement matrix, which will make it possible to predict the strength of such materials, i.e. to create materials with predetermined strength properties.

There are two groups of defects in the structure of cement stone that affect its strength characteristics: rounded like cavities (pores) and pointed like cracks. The pointed ones are decisive in the conditions of material tension, because they cause a high concentration of stresses in the vicinity of vertices, resulting in destruction. Round defects, on the contrary, are quite safe in tension, and under compression conditions initiate tensile stress zones, where cracks appear already under 50% loads of the destructive ones [1]. Cracks and pores with cracks on the material contour are usually randomly oriented relative to the direction of tensile forces. Let's consider the option of placing defects, as given in Fig. 2, which will make it possible to obtain a lower estimate of the material strength, because such a crack orientation is more unfavorable in terms of strength than the stochastic one.

The finite element method is used to calculate stress intensity coefficients (SIC) in a plate with a two-period system of cracks in the holes contour. The results of the calculation are shown in the form of SIC dependence on the parameter which characterizes the distance between the defects in the vertical plane (Fig. 3). As one can see, the two-period system of holes with cracks can be replaced by a collinear one-period system without a significant loss of accuracy in the calculations within the framework of parameter change $g \geq 0{,}3$ *mm*.

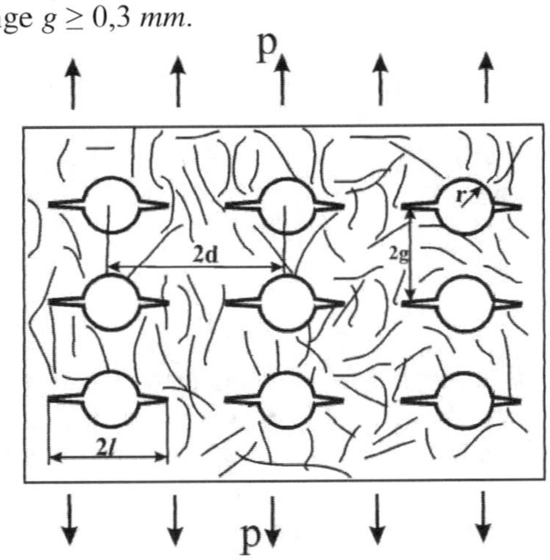

Figure 2. Schematic representation of reinforced cement stone with cracks and pores.

The influence of the size of the holes on SIC in the plate with a two-period system of holes with cracks was also determined by the finite element method (Fig. 4) [9]. It was found that the hole will have little effect on the SIC, if $r/l < 0,7$. Thus, the results of calculations by the finite element method of the problem of a two-period system of holes with cracks on their contour (Fig. 2) give grounds for replacing these defects with a system of collinear cracks with a slight error, which will be assigned to the safety margin. Due to the small size of the defects (0.1… 1 mm), the problem of the periodic system of collinear cracks is considered within the δ_c-model framework (Fig. 5) as more appropriate in this case.

As it is known, Irwin criterion and the concept of stress intensity coefficients are not applicable to minor defects.

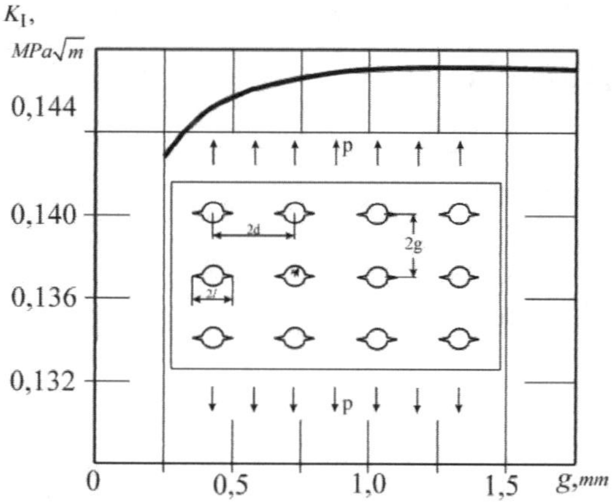

Figure 3. Defects influence on Stress Intensity Factor in parallel planes.

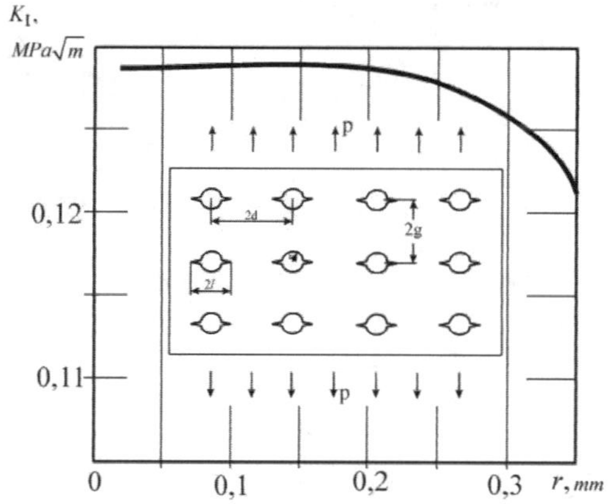

Figure 4. SIC dependence on pores radius.

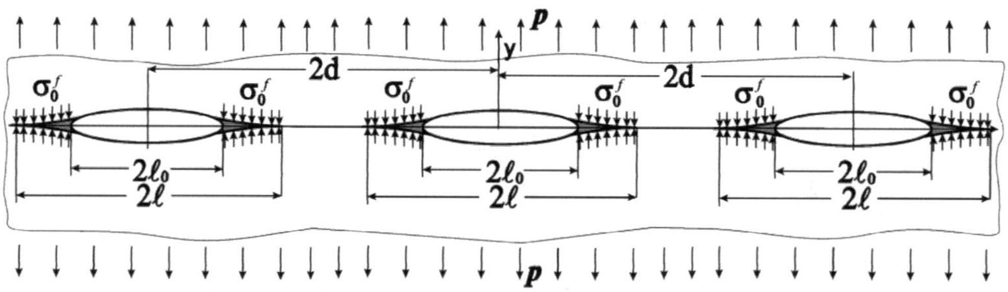

Figure 5. Design diagram of the material with cracks tension.

We use the well-known solution of this problem within the Leonov – Panasyuk model [8, 10], where we obtain a relation that connects the intensity of external load p with the size of the fracture process zone.

$$cos\left(\frac{\pi p}{2\sigma_0}\right) = \frac{\sin\left(\pi a/_{2d}\right)}{\sin\left(\pi l/_{2d}\right)}. \tag{1}$$

The value of σ_0 is equivalent to the strength limit of cement stone volumes between pores and cracks.

Taking the condition of merging the fracture process zones of adjacent cracks as a criterion for destruction of the material, we obtain a formula that allows determining the tensile strength of cement stone under tension.

$$R_{bt}^c = p_* = \sigma_0 (1 - \frac{a}{d}). \tag{2}$$

We should note that the ratio of the crack size to the distance between them a/d reflects the content of defects in the material, that is, its damage.

Let's consider now the cement stone in terms of tension reinforced with microfibers of arbitrary stiffness. The thickness of these fibers will be assumed as much smaller than the characteristic size of the dominant defects – pores and cracks. This assumption makes it possible to consider the material between the pores as homogeneous with certain effective characteristics. Reinforcement with continuous or discrete fibers can be both directional and chaotic. The restraining effect of microfibers in the fracture process zones will be expressed by evenly distributed forces σ_0^f. Their intensity is described by the known Kelly – Tyson dependence [11], modified for arbitrary (directed or chaotic) disperse reinforcement.

$$\sigma_0^f = \lambda\sigma_f V_f \left(1 - \frac{l_c}{2l}\right) + \sigma_0 (1 - V_f), \tag{3}$$

where σf – the fibers strength; lc – boundary inefficient fiber length; l – the fiber length; Vf – volume content of fiber; λ – the coefficient of chaotic reinforcement reduction to directional one [12] (λ = 1 – for directional reinforcement along the tensile axis; λ = 0.33 – for chaotic reinforcement).

By accepting into account the dependences (2), (3), we obtain a calculation formula for predicting the composite strength (fiber concrete) based on the cement matrix.

$$R_{bt}^f = \left(1 - \frac{a}{d}\right)\left(\lambda\sigma_f V_f \left(1 - \frac{l_c}{2l}\right) + \sigma_0 (1 - V_f)\right). \tag{4}$$

Figure 6 shows the graphical dependences of the materials R_{bt}^f strength with different volumetric contents of basalt, polypropylene, glass and carbon fibers and damage to the cement matrix. The mechanical characteristics of the fibers are provided in table 1.

Table 1. Fiber mechanical properties

Fiber name	Density, g/cm3	Tensile strength, MPa	Elasticity modulus, GPa
Basalt	2.65	1200	12
Carbon	2	2000	245
Glass	2.6	1050	75
Steel	7.8	900	200
Polypropylene	0.9	500	5

In order to confirm the validity of the obtained analytical dependence (4), experiments were performed to determine the strength of the cement matrix-based composite for different volumetric content of basalt fiber. The following materials were used to prepare the solutions: basalt fiber; Portland cement PC II/A-3-500; superplasticizer.

Based on the above components, prismatic samples with a size of 40×40×160 mm were made for tensile testing under bending.

The water-cement ratio was assumed as 0.4 with plasticizer Berament TB-1 in the amount of 0.04% by weight of cement. 12 mm long fiber (roving) was taken for studies, the fiber was in the unused condition and has the form of threads bundles. The solution was mixed for 5 min until a homogeneous mixture was obtained, after which it was compacted on a vibrating table to remove air bubbles to reduce the number of pores. All series of testing samples were kept in the laboratory for 27 days.

In fig. 7 a solid line corresponds to the design data, the dots indicate the experimental data. As can be seen, the results correlate well enough. From the above data it follows that the tensile strength of the composite increases more than twice at 2% of the fiber content compared to the base non-reinforced material.

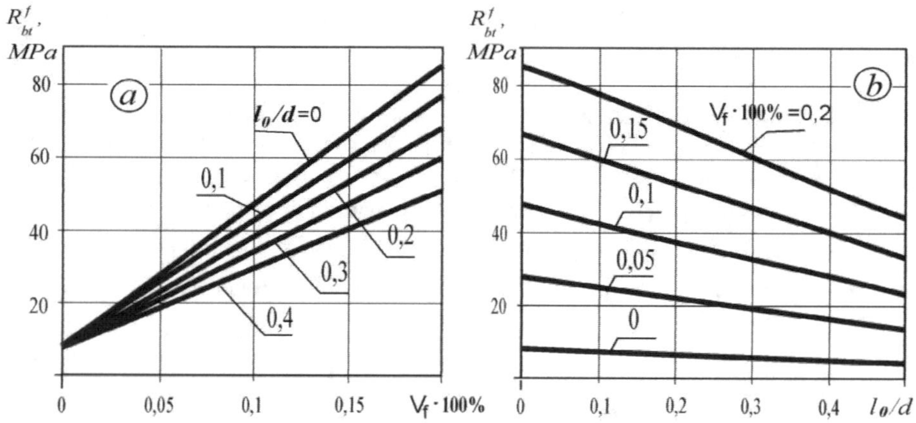

Figure 6. Dependence of the predicted composite strength R^f_{bt} on the volumetric content of reinforcing fibers: carbon (1); basalt (2); fiberglass (3); steel (4) and polypropylene fiber (5); in the absence of pores (a) and the porosity parameter a/d = 0.2 (b).

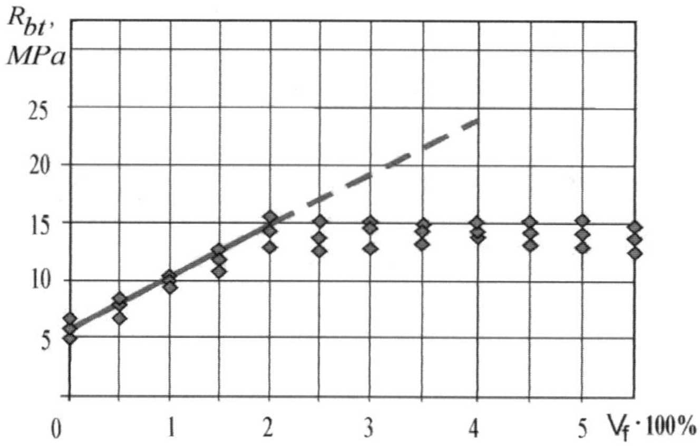

Figure 7.Tensile strength of cement stone reinforced with different basalt fiber content.

Conclusions

Dependence to predict the strength of the cement-based composite matrix tensile fracture mechanics concepts including was obtained. The parameters in the calculation formula are fiber volume content and crack defects, ultimate tensile strength of the composite components, geometric characteristics fiber and others. Calculation on the impact of defects on the strength of cement matrix composite is proposed. Theoretical prediction and experimental strength data correlate very well with each other. This gives reason to use the dependencies when creating composites with desired characteristics on the basis of the concrete matrix.

References

[1] Marukha V Panasyuk V and Sylovanyuk V 2014 *Injection Technologies for Repair of Damaged Concrete Structures* (Springer Dordrecht) p 230 https://doi.org/10.1007/978-94-007-7908-2

[2] Jaworski N Andrushchak N Lobur M Iwaniec M and Sitek W 2019 Composite material microlevel cellular model data transfer and verification by fem analysis software systems *IEEE 15th International Conference on the Perspective Technologies and Methods in MEMS Design (MEMSTECH)* 17-22 doi: 10.1109/MEMSTECH.2019.8817376

[3] Jaworski N Iwaniec M and Lobur M 2019 Implementation features of composite materials effective mechanical characteristics finding method based on microlevel cellular structural

IOP Conf. Series: Materials Science and Engineering 1016 (2021) 012009 doi:10.1088/1757-899X/1016/1/012009

models *15th International Conference on the Experience of Designing and Application of CAD Systems (CADSM)* 1-5 doi: 10.1109/CADSM.2019.8779273

[4] Jaworski N Lobur M and Iwaniec M 2018 Implementation features of cellular composites microlevel structural models construction based on Voronoi tessellation and OpenCL technology usage *14th International Conference on Perspective Technologies and Methods in MEMS Design (MEMSTECH)* 109-13 doi: 10.1109/MEMSTECH.2018.8365713

[5] Jaworski N and Andrushchak N 2017 A method of nanoporous anodic aluminum oxide structure modeling based on Bezier curves generation *14th International Conference The Experience of Designing and Application of CAD Systems in Microelectronics (CADSM)* 63-6 doi: 10.1109/CADSM.2017.7916085

[6] Jaworski N Iwaniec M and Lobur M 2016 Composite materials microlevel structure models visualization distributed subsystem based on WebGL *XII International Conference on Perspective Technologies and Methods in MEMS Design (MEMSTECH)* 22-4 doi: 10.1109/MEMSTECH.2016.7507511

[7] Jaworski N Farmaga I Matviykiv O Lobur M Śpiewak P Ciupiński Ł and Kurzydłowski K J 2014 *Thermal analysis methods for design of composite materials with complex structure ECS Transactions* **59**(1) 513-23 doi: 10.1149/05901.0513ecst

[8] Cherepanov G 1977 *Mechanics of fragile destruction* (Moscow: Nauka) p 640

[9] Nevlyudov I Yevsieiev V Kolesnyk K and Saliieva V 2018 Finite element type and the features of the i-beam simulation influence on the calculation results *CAD in Machinery Design: Implementation and Educational Issues (CADMD)* 57-60

[10] Vytvyzkyj P 1970 Elastic-plastic equilibrium of a plate with a periodic system of slits *AN URSR Ser A* **6** 524-7

[11] Kelly A and Tyson W 1965 Fiber Strengthened Materials *in High Strength Materials* (New York: John Wiley & Sons Inc) 578-83

[12] Rabinovych F 2004 Composites based on dispersed reinforced concrete *Problems of theory and design, technologies, structures* 560

IOP Conf. Series: Materials Science and Engineering 1016 (2021) 012010 doi:10.1088/1757-899X/1016/1/012010

Modeling Issues in Problems of the Elasticity and Viscoelasticity Theory

T Kagadiy [1], A Shporta [1], I Scherbina[2]* and O Onopriienko[3]

[1]National TU Dnipro Polytechnic, Department of Higher Mathematics, 49005, Yavornytsky D. avenue, 19, Dnipro, UKRAINE
[2]Dnipro State Agrarian and Economic University, Department of Higher Mathematics and Physics 49000, S. Efremova S., 25, Dnipro, UKRAINE
[3]Dnipro State Agrarian and Economic University, Department of Theoretical Mechanics 49000, S. Efremova S., 25, Dnipro, UKRAINE

*Email: sherbinaiv@ukr.net

Abstract. The development of computer technology and information technology does not exclude preliminary analytical modeling in complex problems of mechanics. The approach proposed in this work allows one to obtain reasonable simplified equations that admit analytical or numerical-analytical solutions. Authors research are devoted to method elaboration for space problems of viscoelasticity. The approach for solving problems of nonlinear elasticity theory, when final deformations or physical nonlinear properties of materials are taken into consideration, is proposed. New perturbation method for solving of nonlinear equations in particular derivatives is suggested. Such approach is allowed to reduce the solution of complicated problems of linear elasticity and viscoelasticity to subsequently solved boundary problems of potential theory. New linear problems are investigated, in particular, problems on load transference from stringer to single layer or multilayer solids, on stress-strain state of fibrous composite with crack in matrix (plate and axisymmetric problems), numerous contact problems (pressing of hard stamp in an orthotropic plate with different anisotropy). Problems of nonlinear elasticity theory on stress-strain state of a plate with a circular hole under different types of loading are solved.

1. Introduction.

The problems of transferring loads from reinforcing elements to bodies with different properties are directly related to structural mechanics. Some issues of fracture of fiber-reinforced composites can also be considered. In this work, the formulation of the problem is complicated by taking into account the viscoelastic properties of the base material, which consists of two layers. Analytical solutions for spatial problems of this type are almost never found in the literature. Therefore, this investigation is actual and can be useful for further numerical calculations. The works of many scientists are devoted to the methods of solving plane and axisymmetric contact problems on the transfer of loads from stamps, linings and other reinforcing elements to bases with different properties [1, 2, 7].

The aim of research in recent years is to take into account the complex properties of materials, which brings the mathematical model closer to real problems. For example, in [3], a method was proposed for solving the problem of electroelasticity for multiply connected plates.

IOP Conf. Series: Materials Science and Engineering 1016 (2021) 012010 doi:10.1088/1757-899X/1016/1/012010

2. Building a mathematical model

Consider an axisymmetric contact problem on the transfer of a load from a rod of circular cross-section to a viscoelastic body, which consists of two orthotropic layers with cylindrical anisotropy attached to each other

$$0 \le z_1 \le h_1, h_1 \le z_2 \le h_2, r \ne \infty, z_1 = z, z_2 = z - h_1.$$

The rod is located perpendicular to the bounding plane, its middle line coincides with the axis Oz. It is required to find the distribution law of contact stresses between the stringer and the body, as well as the forces in the stringer under the condition of its loading at the end points by longitudinal forces $P_0^{(1)}$ and $P_0^{(2)}$ (indices 1, 2 refer to the corresponding layers). At finite values h_1, h_2 from the equilibrium conditions $P_0^{(1)} = P_0^{(2)}$.

Since in spatial problems for bodies with inclusions, the model of a one-dimensional elastic bar together with a contact model along a line is not used, it is assumed that there is a model of a one-dimensional inclusion in combination with a contact model along a cylindrical surface for the base.

Since the base material is viscoelastic with cylindrical anisotropy, and the problem is formulated taking into account the axial symmetry of the loading, then the stress tensor and displacement vector do not depend on Q (rQz cylindrical coordinates).

In this case, the problem is divided into two independent ones: the deformation problem, in which one displacement component is missing, and the torsion problem. The first of them is considered in detail. After applying the Laplace transform in time to the relationship between strains and stresses, the problem is reduced to integrating the equilibrium equations with respect to transformants for each layer [5]

$$\frac{\partial^2 u}{\partial r^2} + \varepsilon_1 \frac{\partial^2 u}{\partial z^2} + \varepsilon_1 m \frac{\partial^2 \varpi}{\partial r \partial z} + \frac{1}{r}\frac{\partial u}{\partial r} - \frac{u}{r^2} = 0$$

$$\varepsilon_1 \frac{\partial^2 \varpi}{\partial r^2} + \frac{\partial^2 \varpi}{\partial z^2} + \varepsilon_1 m \frac{\partial^2 u}{\partial r \partial z} + \varepsilon_1 \frac{1}{r}\frac{\partial \varpi}{\partial r} + \varepsilon_1 m \frac{1}{r}\frac{\partial u}{\partial z} = 0 \qquad (1)$$

$$\varepsilon_1 = \varepsilon \Phi(p) \quad \varepsilon = G/E_1$$

under the following boundary conditions

$$\sigma_{33} = E_3\left(\frac{\partial \varpi}{\partial z} + \nu_{31}\frac{\partial u}{\partial r} + \nu_{32}\frac{u}{r}\right) \sigma_{13} = G\left(\frac{\partial \varpi}{\partial r} + \frac{\partial u}{\partial z}\right), \sigma_{33} = \sigma_{13} = 0 \ (z = 0, z = h_1 + h_2)$$

$\varpi = \varpi_c, u = 0 \quad (r = a)$

When $z = h_1$ the layers transformants of displacement are equal to each other. The same condition is fulfilled for stresses σ_{33}, σ_{13}. All functions turns to zero at infinity. The displacement of the bar ϖ_c satisfies the relations

$$E_c F_c d^2\varpi_c / dz^2 = q(z) \qquad N = E_c F_c d\varpi_c / dz = P_0 \qquad (z = 0, z = h_1 + h_2)$$

Here u, ϖ_c are the transformants of the components of the displacement vector of the corresponding layers, E_1, E_2, E_3 (G) are analogs of the elasticity (shear) moduli of the materials of the layers, taking into account the functions that appear when taking into account the viscoelasticity, $m = G + \nu_{13}E_1$.

The value ε_1 is considered as a "small" parameter [4] in the asymptotic integration of the system (1). It is really small, since ε it is a small value, and the function $\Phi(p)$ does not exceed unity for the difference creep kernels (p is the Laplace transform parameter). $q(z) = -2\pi a \sigma_{13}(a, z)$ - the force of contact interaction between the bar and the layers, a is the radius of the bar. Since at $r = a$ the displacement transformant u and the derivative $\partial u/\partial z$ are equal to zero, then the stresses $\sigma_{13}(a,z)$ are completely determined only by the function $\partial\varpi/\partial z$:

$$q(z) = -2\pi a G(\partial\varpi/\partial r)\big|_{r=a}$$

To determine the effort in the stringer and the effort of contact interaction, the method proposed by the authors is used [4,5]. After the decomposition of the stress-strain state into two components, the boundary value problem is reduced to the following for the first layer:

$$E_3^1\frac{\partial^2\varpi_1}{\partial z^2} + G^{(1)}\left(\frac{\partial^2\varpi_1}{\partial r^2} + \frac{1}{r}\frac{\partial\varpi_1}{\partial r}\right) = 0 \tag{2}$$

$$\text{for } z=0, \frac{\partial\varpi_1}{\partial z} = 0, \quad \text{for } z = h_1 \quad \frac{\partial\varpi_1}{\partial z} = \frac{f(r)}{E_3^{(1)}} \quad \text{for } r=a \quad \varpi_1 = \varpi_c \quad \left(\frac{\partial\varpi_1}{\partial z}\Big|_{r=a} = \frac{\partial\varpi_c}{\partial z}\right)$$

$$E_cF_c d^2\varpi_c/dz^2 = -2\pi a G^1(\partial\varpi_1/\partial r)\big|_{r=a} \tag{3}$$

$$\text{for } z=0, \frac{\partial\varpi_c}{\partial z} = \frac{P_0^{(1)}}{E_cF_c}, \text{ for } z = h_1 \quad \frac{\partial\varpi_c}{\partial z} = \frac{\partial\varpi_1}{\partial z}\Big|_{r=a} = \frac{f(a)}{E_3^{(1)}}$$

For the second layer, similarly.

Note once again that the writing of equations through transformants coincides with the corresponding equations for an elastic material, since the additional functions that arise due to viscoelasticity are introduced into the coefficients (constants) in the equations. After applying to equations (2), (3) the Fourier cosine transform in the coordinate z with finite limits, we obtain

$$\frac{d^2\varpi_1^*}{dr^2} + \frac{1}{r}\frac{d\varpi_1^*}{dr} - \varpi_1^2\left(\frac{n\pi}{h_1}\right)\varpi_1^*(r,n) = -\frac{(-1)^n}{G^{(1)}}f(r)$$

$$E_cF_c\left(\frac{n\pi}{h_1}\right)^2\varpi_1^*(a,n) = \frac{E_cF_c}{E^{(1)}}f(a)(-1)^n - p_0^{(1)} + 2\pi a G^{(1)}\frac{d\varpi_1^*}{dr}\Big|_{r=a} \tag{4}$$

$$\varpi_1^*(r,n) = \int_0^{h_1}\varpi_1(r,z)\cos\frac{n\pi z}{h_1}dz, \quad \varpi_1^2 = E_3^{(1)}/G^{(1)} \quad . \tag{5}$$

After applying the Weber transforms with respect to the coordinate r to equations (4) taking into account the boundary conditions, we obtain

$$(2/\pi)\varpi_1^*(a,n) - \lambda^2 W_1(\lambda,n) - \varpi_1^2(n\pi/h_1)W_1(\lambda,n) = -\left((-1)^n/G_1\right)F(\lambda), \quad F(\lambda) = \int_a^\infty f(r)r\phi\lambda(r)dr,$$

$$W_1(\lambda,n) = \int_a^\infty \varpi_1^*(r,n)r\phi\lambda(r)dr, \quad \phi_\lambda(r) = I_0(a\lambda)Y_0(\lambda r) - Y_0(a\lambda)I_0(\lambda r)$$

I_v, Y_v - Bessel functions of the first and second kind.

After applying the inverse Weber transform and calculating some integrals, the need for the inverse Fourier transform remains. Dirichlet's theorem is used for this [6]. As a result, we get

$$\varpi_1(a,z) = \frac{2h_1}{\pi^2}\sum_{n=1}^\infty\left(\frac{f(a)}{E_3^{(1)}}(-1)^n - \frac{P_0^{(1)}}{E_cF_c} + \frac{2\pi}{E_cF_c}(-1)^n A_1(n)\right)\frac{\cos(n\pi z/h_1)}{n(n+g_1M_1(n))}$$

$$\frac{d\varpi_1}{dr}\Big|_{r=a} = \frac{2}{G^{(1)}ah_1}\sum_{n=1}^\infty(-1)^n A_1(n)\cos(n\pi z/h_1) -$$

$$-\frac{2\varpi_1}{\pi}\sum_{n=1}^{\infty}\left(\frac{f(a)}{E_3^{(1)}}(-1)^n-\frac{P_0^{(1)}}{E_cF_c}+\frac{2\pi}{E_cF_c}(-1)^n A_1(n)\right)\frac{\cos(n\pi z/h_1)}{nM_1^{-1}(n)+g_1}$$

$$f(r)=\int_0^{\infty}\left(F(\lambda)\phi_\lambda(r)\lambda d\lambda\right)/\left(J_0^2(a\lambda)+Y_0^2(a\lambda)\right)$$

For the second layer, the solution is carried out similarly, taking into account the change of variable $z-h_1=\xi$ in the equations(3), (4).

3. Main results and their discussions

After estimating a number of integrals using the mean value theorem and taking into account the boundary conditions, we find the sought functions and their derivatives for two layers

$$w_1(a,z)=\frac{2h_1}{\pi^2}\frac{f(a)}{E_3^{(1)}}\sum_{n=1}^{\infty}\frac{(-1)^n\cos(n\pi z/h_1)}{n(n+g_1M_1(n))}-\frac{2h_1}{\pi^2}\frac{P_0^{(1)}}{E_cF_c}\sum_{n=1}^{\infty}\frac{\cos(n\pi z/h_1)}{n(n+g_1M_1(n))}$$

$$w_2(a,\xi)=\frac{2h_2}{\pi^2}\frac{P_0^{(2)}}{E_cF_c}\sum_{n=1}^{\infty}\frac{(-1)^n\cos(n\pi\xi/h_2)}{n(n+g_2M_2(n))}-\frac{2h_2}{\pi^2}\frac{f(a)}{E_3^{(2)}}\sum_{n=1}^{\infty}\frac{\cos(n\pi\xi/h_2)}{n(n+g_2M_2(n))}$$

$$\frac{dw_1(a,z)}{dz}=-\frac{2}{\pi}\frac{f(a)}{E_3^{(1)}}\sum_{n=1}^{\infty}\frac{(-1)^n\sin(n\pi z/h_1)}{(n+g_1M_1(n))}-\frac{2}{\pi}\frac{P_0^{(1)}}{E_cF_c}\sum_{n=1}^{\infty}\frac{\sin(n\pi z/h_1)}{(n+g_1M_1(n))}$$

$$\frac{dw_2(a,\xi)}{dz}=-\frac{2}{\pi}\frac{P_0^{(2)}}{E_cF_c}\sum_{n=1}^{\infty}\frac{(-1)^n\sin(n\pi\xi/h_2)}{(n+g_2M_2(n))}+\frac{2}{\pi}\frac{f(a)}{E_3^{(2)}}\sum_{n=1}^{\infty}\frac{\sin(n\pi\xi/h_2)}{(n+g_2M_2(n))}$$

$$\frac{\partial w_1}{\partial r}\Big|_{r=a}=-\frac{2\omega_1}{\pi}\frac{f(a)}{E_3^{(1)}}\sum_{n=1}^{\infty}\frac{(-1)^n\cos(n\pi z/h_1)}{nM_1^{-1}(n)+g_1}+\frac{2\omega_1}{\pi}\frac{P_0^{(1)}}{E_cF_c}\sum_{n=1}^{\infty}\frac{\cos(n\pi z/h_1)}{nM_1^{-1}(n)+g_1}$$

$$\frac{\partial w_2}{\partial r}\Big|_{r=a}=-\frac{2\omega_2}{\pi}\frac{P_0^{(2)}}{E_cF_c}\sum_{n=1}^{\infty}\frac{(-1)^n\cos(n\pi\xi/h_2)}{nM_2^{-1}(n)+g_2}+\frac{2\omega_2}{\pi}\frac{f(a)}{E_3^{(2)}}\sum_{n=1}^{\infty}\frac{\cos(n\pi\xi/h_2)}{nM_2^{-1}(n)+g_2}$$

Unknown value $f(a)$ is found from condition $w_1(a_1,h_1)=w_2(a,0)$.

$$f(a)=\frac{P_0^{(1)}E_3^1}{E_cF_c}\left(\sum_{n=1}^{\infty}\frac{(-1)^n}{n(n+g_1M_1(n))}+\frac{h_2}{h_1}\frac{P_0^{(2)}}{P_0^{(1)}}\sum_{n=1}^{\infty}\frac{(-1)^n}{n(n+g_2M_2(n))}\right)/$$
$$\left(\sum_{n=1}^{\infty}\frac{(-1)^n}{n(n+g_1M_1(n))}+\frac{h_2}{h_1}\frac{E_3^{(1)}}{E_3^{(2)}}\sum_{n=1}^{\infty}\frac{(-1)^n}{n(n+g_2M_2(n))}\right).$$

The load in the stringer $N(z)$ and the load of contact interaction are according to the formulas $N(z)=E_cF_cdw(a,z)/dz, \quad q(z)=-2\pi aG(\partial\varpi/\partial r)\big|_{r=a}$

If $h_2=0$ then we come to the problem of transferring the load from the bar to the thickness h_1 layer. If $h_2=0,\ h_1\to\infty$ then we obtain a solution to the problem of transferring a load to an orthotropic semi-infinite viscoelastic body. Such passage to the limit allows you to check the results obtained by the proposed method, since the simplified versions of the problem have already been solved.

The inverse Laplace transform determines N and q depending on the coordinates and time. To go to the originals, the effort is presented in the form of rows in a small parameter $\bar{\varepsilon}$ which depends on p

$$N(z,p) = \left(N_0(z) + N_1(z)\bar{\varepsilon} + N_2(z)\bar{\varepsilon}^2 + \ldots\right)/p$$
$$q(z,p) = \left(q_0(z) + q_1(z)\bar{\varepsilon} + q_2(z)\bar{\varepsilon}^2 + \ldots\right)/p \qquad (6)$$

The coefficients of these expansions are easier to find for "small" and "large" values of the parameter. Such limiting values of functions can be connected using the Padé approximant [5].

For example, in the problem of transferring a load to a half-space, which has predominantly shear creep at $p \to \infty$, $\bar{\varepsilon} = k/(p+\beta)$, for $p \to 0, \bar{\varepsilon} = \Delta\, p/(p+\beta)$, $\Delta = -k/(\beta+k)$.

The originals of the functions for small values of time have the form

$$N(z,t) = N_0(z) + N_{10}(z)(K/\beta)(1-e^{-\beta t}) + \ldots,$$

$$N_0(z) = \frac{2P_0}{\pi}\int_0^\infty \frac{\sin zs}{s + g_0 M_0(s)}ds,$$

$$N_{10}(z) = \frac{2P_0}{\pi}\int_0^\infty \frac{\left[g_0 a\omega_0 s\left(1-M_0^2(s)\right) + 2g_0 M_0(s)\right]\sin zs}{2\left[s+g_0 M_0(s)\right]^2}ds,$$

$$q_0(z) = \frac{2P_0 g_0}{\pi}\int_0^\infty \frac{\cos zs}{g_0 + sM_0^{-1}(s)}ds, \qquad (7)$$

$$q_{10}(z) = \frac{2P_0 g_0}{\pi}\int_0^\infty \frac{\left[a\omega_0 s^2\left(1-M_0^{-2}(s)\right) - 2sM_0^{-1}(s)\right]\cos zs}{2\left[g_0 + sM_0^{-1}(s)\right]^2}ds,$$

$$M_0(s) = K_1(a\omega_0 s)/K_0(a\omega_0 s).$$

At large times

$$N(z,t) = N_\infty(z) + N_{1\infty}(z)\Delta e^{-\chi t} + \ldots$$
$$q(z,t) = q_\infty(z) + q_{1\infty}(z)\Delta e^{-\chi t} + \ldots$$

The expansion coefficients are found from (7) after replacing ω_0 by ω_∞, g_0 by g_∞ for

$$\omega_\infty = \omega_0(1+k/\beta)^{1/2}, \quad g_\infty = g_0(1+k/\beta)^{-1/2}.$$

For the same case, one of the Padé approximants has the form

$$N^* = \frac{0.915 - 0.11t - 1.31e^{0.5t}}{1 - 0.134t - 1.55e^{0.5t}}.$$

Figures 1, 2 show the limiting values $N^* = N/P_0$ and $q^* = q/P_0 g_0$ at $t \to 0$ (curve 1) and $t \to \infty$ (curve 2).

IOP Conf. Series: Materials Science and Engineering 1016 (2021) 012010 doi:10.1088/1757-899X/1016/1/012010

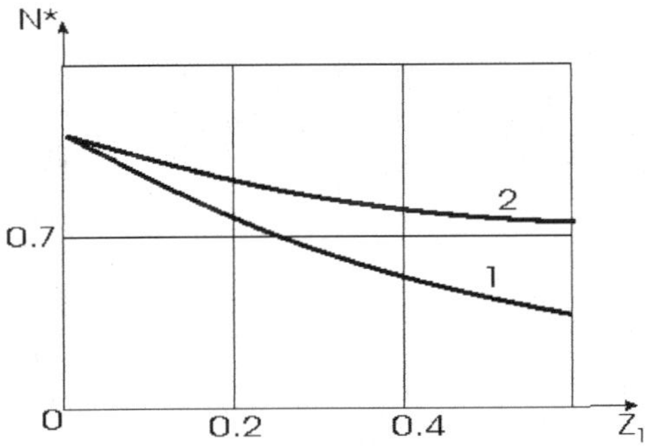

Figure 1. The core tension change

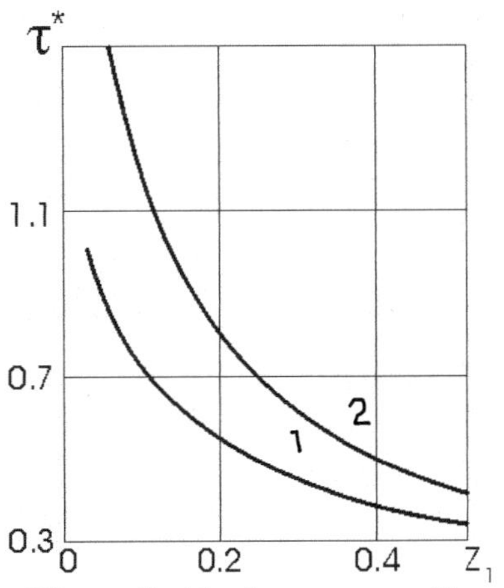

Figure 2. Altering contact efforts

Conclusions

The practical significance of the obtained results lies in the possibility of using the proposed approach to construct reasonable simplified boundary value problems that bring the model as close as possible to the real problem. The results can be used for numerical implementation in the design of complex multi-layer structures from modern materials. The obtained solutions of the spatial contact problem can be used as exact solutions in the estimation of approximate computations.

References

[1] Guz A N Babich S Yu and Rudnitskiy V B 2013 Kontaktnoe vzaimodeystvie uprugih tel s nachalnyimi (ostatochnyimi) napryazheniyami. Razvitie idey L. A. Galina v mehanike *(Moskva Izhevsk Izd-vo In-ta kompyuternyih issledovaniy* 2013) 480

[2] Pryvarnykov A K and Spytsia O H 2005 Osesymetrychni kontaktni zadachi dlia pruzhnykh bahatosharovykh plyt *Visnyk Donetskoho universytetu Seriia A Pryrodnychi nauky* **1(1)** 53-7

[3] Kaloerov S A and Samodurov A A 2014 Zadacha elektrovyazkouprugosti dlya mnogosvyaznyih plastinok *Matematichni metodi ta fIziko-mehanichni polya* **57(3)** 62–77

[4] Kagadiy T S and Shporta A H 2015 The asymptotic method in problems of the linear and nonlinear elasticity theory *Naukovyi Visnyk Natsionalnoho Hirnychoho Universytetu (Dnipro)* **3** 76-81

[5] Kahadiy T S Belova O V and Shcherbina I V 2019 Zastosuvannia metodu maloho parametru pry modeliuvanni zadach teorii v′iazkopruzhnosti *Visnyk Khersonskoho natsionalnoho universytetu (Kherson)* **2(69)** 69-76

[6] Kagadiy T S Shporta A H Bilova O V and Scherbina I V 2020 Stressed-strained state of a layered basis with a fastening element *Applied questions of mathematical modeling* **3(2.1)** 107-16 https://doi.org/10.32782/KNTU2618-0340/2020.3.2-1.10

[7] Kagadiy T and Shporta A 2019 Mathematical modeling in the calculation of reinforcing element *Naukovyi Visnyk Natsionalnoho Hirnychoho Universytetu (Dnipro)* **5** 60-4 https://doi.org/10.29202/nvngu/2019-5/10

IOP Conf. Series: Materials Science and Engineering 1016 (2021) 012011 doi:10.1088/1757-899X/1016/1/012011

Experimental Model of a Mobile Platform for Moving on Orthogonal Routes

Vitaliy V Mazur* and Sofiia T Panchak

CAD Systems Department, Lviv Polytechnic National University, Lviv, 5 Mytropolyta Andreya, 79013, Ukraine

*Email: mzvitg@gmail.com

Abstract. The paper considers an experimental model of a mobile platform for moving on orthogonal routes. The advantages of orthogonal routes for planning and ensuring the movement of the platform are shown. The structural scheme of the mobile platform and the organization of functioning of its subsystems are developed. The platform's construction and digital control system for ensuring its motion and positioning on the marked surface are offered. Methods of traffic organization and routes planning for platform movement are proposed. Peculiarities of using the experimental model of the platform in the educational process of the CAD Department are considered.

1. Introduction

Mobile platforms are an important component of modern robotic systems. Many theoretical works and practical developments are devoted to various aspects of creating such platforms [1, 2]. One of the promising areas that is currently developing intensively is the further intellectualization of mobile platforms, which ensures their movement along arbitrary trajectories [3-6]. In some cases, this expands the capabilities of such platforms, but leads to further complication of control systems and an increase in their cost. In addition, when using several platforms on the same route network, the task of controlling their joint movement in a limited space becomes much more complicated due to mutual interference [7]. At the same time, for solving a number of practical problems, the introduction of certain restrictions is quite acceptable, which makes such platforms simpler and cheaper. One of the possibilities for introducing such restrictions in this work is the transition from movement along arbitrary routes to the movement of platforms on orthogonal routes. This reduces the cost of the mobile platform control system and simplifies the routes planning and synchronization of mobile platforms movement when they move together. In our opinion, some aspects of creating mobile platforms for moving on orthogonal routes and planning such routes have not been sufficiently studied. This determines the relevance of such research. Therefore, the problem of creating an experimental model of a mobile platform for researching its movement on orthogonal routes is posed and solved in this work.

The purpose of the work: Development of the structure, construction and control system of the experimental model of a mobile platform for movement on orthogonal routes.

To achieve this goal, the following tasks were solved in the work.

- The development of the kinematic scheme and construction of the platform.
- The development of the general structure of the mobile platform control system and definition of its main components.

- The development of methods and control systems to ensure the movement of the platform on orthogonal routes.
- Development of methods of route planning on an orthogonal route network.

2. The development of the kinematic scheme and construction of the platform

The main task which is solved by the developed mobile platform is to provide controlled movement on the planned route. To do this, it is proposed to use an orthogonal route network, which provides deterministic movement of the mobile platform, simplifies route planning and provides for a discrete change of direction at intersections. The trajectory of the mobile platform is determined by the marking on the traffic surface, which determines the configuration of the route network. The construction of the mobile platform includes a mechanical part that implements the kinematic scheme of the platform and an electronic part that implements a digital motion control system of the platform. The experimental model of the mobile platform placed on the marked surface is presented in Figure 1.

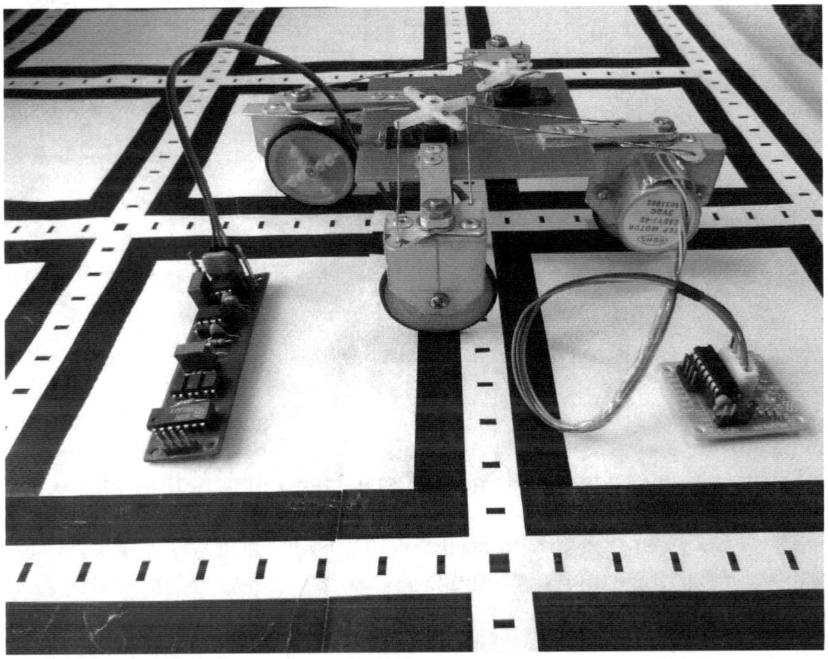

Figure 1. Experimental model of a mobile platform on a marked surface.

The proposed kinematic scheme of the platform is presented in Figure 2. It consists of two independent drive units equipped with controlled servo drives (SG92R), to change the direction of the platform movement. Digitally controlled servo drives provide synchronous turn of drive and passive wheels by ±90 degree when changing the direction of the platform movement, as well as turn of these wheels from the main direction by ±15 degree to compensate the deviations of the platform from the trajectory specified by the marking lines. The drive wheel of each drive unit is equipped with the stepper motor with a reducer (28BYJ-48) and provides the controlled movement of the platform on a route with the set speed. The wheel layout of each unit provides independent displacement of the front and rear of the platform based on signals from optical sensors that track the marking lines. Changing the direction of movement by ±90 degree is carried out for a motionless platform at the intersections of the route network. One of the advantages of the proposed kinematic scheme is that platform orientation does not change when changing the direction of platform movement at 90 degree. That is, the coordinate system of the platform coincides with the coordinate system of the plane on which it moves. This simplifies the system of external control of the manipulator, which a mobile platform can be equipped.

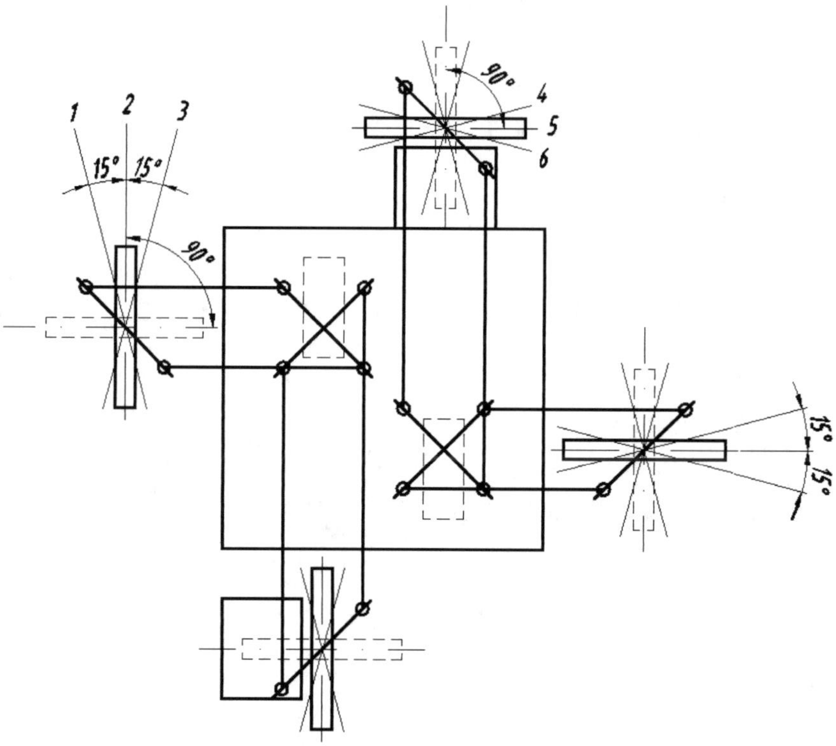

Figure 2. Kinematic scheme of the mobile platform.

3. The development of the general structure of the mobile platform control system

The digital control system for the mobile platform provides the solution of two main tasks: change of the direction of the platform movement and controlled movement of the platform on the certain distance with the set speed. Changing the direction of the platform movement includes changing the main direction of movement by ±90 degrees at intersections and periodically changing the direction of movement by ±15 degrees from the main direction to automatically track marking lines and correct deviations from the trajectory specified by the route. The general structure of the mobile platform control system is presented in Figure 3. It consists of two subsystems: the subsystem for controlling the change of movement direction and the subsystem for controlling the movement of the platform. Changing the direction of movement of the front and rear of the platform is carried out using servo drives (SD), which change the angle of rotation of the wheels by ±90 degrees or ±15 degrees. The formation of control pulses for servo drives is provided by control pulses shaper for servos (CPSS) using pulse-width modulation (PWM). The formation of deviation signals from the marking lines is carried out by the sensors unit (SU). The sensors signal switch (SSS) switches groups of signals depending on the direction of the platform movement. Control signals to change the direction of movement by ±90 degrees and ±15 degrees are generated by the microcontroller based on the route description and signals from the sensors unit. The platform movement control subsystem consists of two driving stepper motors (SM). The formation of the pulse sequence for the half-step mode of stepper motors operation is carried out by control pulses shaper for stepper motors (CPSSM). The speed and direction of rotation the shaft of stepper motors is determined by the control signals coming from the microcontroller (MCU). The generated pulses are fed to the stepper motors drivers (SMD). Centralized route planning for one or more mobile platforms is provided by a personal computer (PC), which transmits the necessary information to the MCU. For the experimental model of the mobile platform the personal computer performs the functions of the MCU.

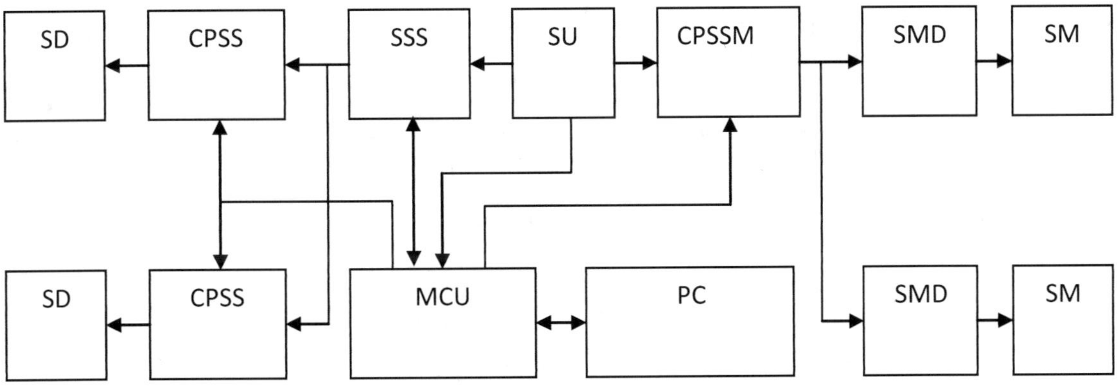

Figure 3. The general structure of the mobile platform control system.

4. Control system for changing the direction of the platform movement and tracking a given trajectory

Changing the direction of movement of the platform and tracking a given trajectory is carried out by servo drives with digital control. Each servo provides synchronous turn of the drive and passive wheel by ±90 degrees to change the moving direction of the platform at the intersection of routes and change the angle of rotation of the wheels by ±15 degrees from the main direction to compensate the deviations from the specified trajectory during the platform movement along the route. The electrical diagram of the control system is shown in Figure 4. Pulse width modulation (PWM) is used to control the servo. The proposed control circuit consists of a generator of synchropulses with a period of 20 milliseconds, implemented as a multivibrator based on the timer DA1 NE 555. The synchropulses are fed to a pulse-width modulator implemented on the basis of a singlevibrator (chip DA2 NE 555), which generates pulses with a given duration. These pulses are fed to the control input of the servo. The pulse duration is determined by the time constant of the RC-link formed by the capacitor C3 and a set of resistors R7-R10. Resistors R7, R8 and R9 shunted by diode-transistor optocouplers (EL817) to change the time constant and generate control pulses of different duration. The optocoupler are controlled by digital input signals K_1, K_2, K_3. The relationship between the angle of rotation of the servo and the duration of the control pulses is shown in Figure 5. Provision of 12 necessary directions of movement is carried out by changing the angle of rotation of the wheels (signals K_1, K_2, K_3) and the direction of the stepper motors rotation (signal D). The values of the signals to ensure the specified direction of movement N are presented in the correspondence table 1.

The reflective optical sensors system S1-S8 (TCRT5000) is used to detect the deviations of the mobile platform from the trajectory set by the black and white marking lines (Figure 6). Signals from optosensors are fed to the Schmidt trigger, which improves the shape of digital signals and provides the hysteresis needed to eliminate self-oscillations of the tracking system at the boundary of the black and white marking lines. When moving in the direction, indicated by the arrow, sensors S1 and S2 detect the deviation of the front part of the platform (to left or to right), and sensors S3 and S4 detect the deviation of the rear part of the platform. Based on the signals S_1-S_4 from the optosensors S1-S4, taking into account the direction of rotation of the stepper motor D, the signals K_1-K_3 are determined (in accordance with table 2) to generate width-modulated pulses. These pulses control the servos SD1 and SD2. When moving in the perpendicular direction, switching to sensors S5-S8 is performed by sensors signal switch SSS. The values of the signals K_1-K_3, which provide the formation of width-modulated pulses for servos SD1 and SD2 depending on the signals S_5-S_8 of the optosensors S5-S8 and the direction of rotation of the stepper motor D, are presented in table 3. Automatic tracking of marking lines ensures the coincidence of the coordinate system of the mobile platform and the marked traffic surface.

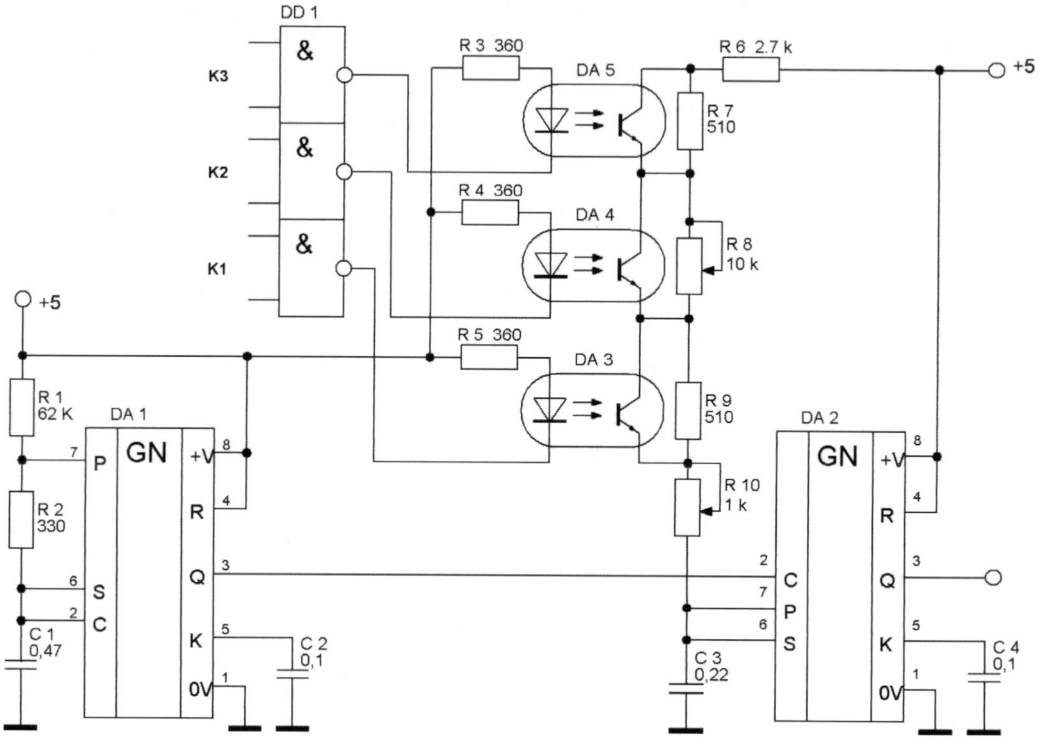

Figure 4. Electrical circuit for the formation of the servo control pulses.

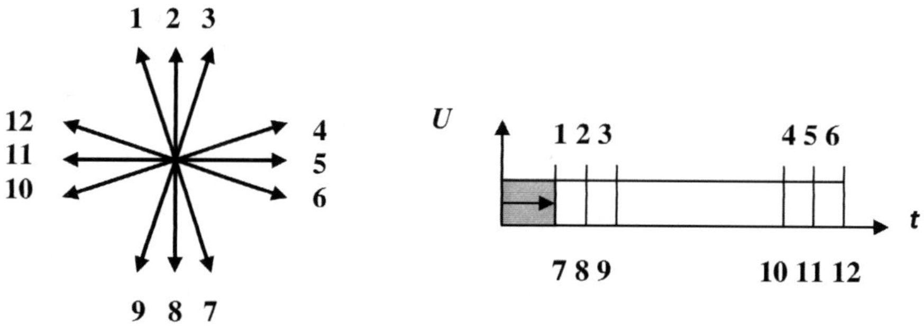

Figure 5. The width of the control pulse of the servo to provide the necessary directions of movement.

Table 1. Control signals to provide the required directions

N	D	K_1	K_2	K_3	N	D	K_1	K_2	K_3
1	0	1	1	1	7	1	1	1	1
2	0	0	1	1	8	1	0	1	1
3	0	0	1	0	9	1	0	1	0
4	0	1	0	1	10	1	1	0	1
5	0	0	0	1	11	1	0	0	1
6	0	0	0	0	12	1	0	0	0

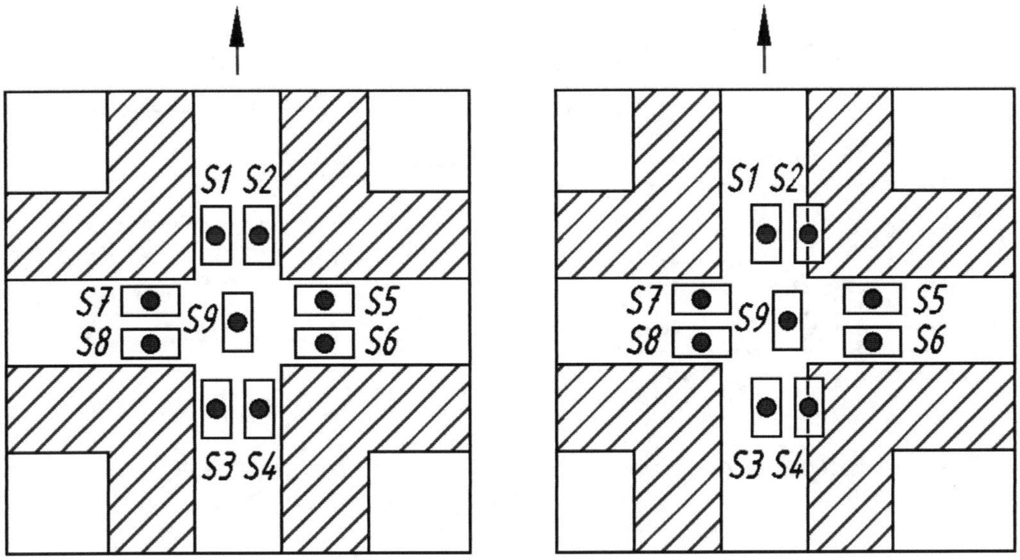

Figure 6. Reflective optical sensors system and its functioning

Table 2. Control signals to provide the required directions

S_1	S_2	D	K_1	K_2	K_3	N	SD	S_3	S_4	D	K_1	K_2	K_3	N	SD
1	1	0	0	1	1	2	1	1	1	0	0	1	1	2	2
0	1	0	0	1	0	3	1	0	1	0	0	1	0	3	2
1	0	0	1	1	1	1	1	1	0	0	1	1	1	1	2
1	1	1	0	1	1	8	1	1	1	1	0	1	1	8	2
0	1	1	1	1	1	7	1	0	1	1	1	1	1	7	2
1	0	1	0	1	0	9	1	1	0	1	0	1	0	9	2

Table 3. Control signals to provide the required directions

S_5	S_6	D	K_1	K_2	K_3	N	SD	S_7	S_8	D	K_1	K_2	K_3	N	SD
1	1	0	0	0	1	5	1	1	1	0	0	0	1	5	2
0	1	0	0	0	0	6	1	0	1	0	0	0	0	6	2
1	0	0	1	0	1	4	1	1	0	0	1	0	1	4	2
1	1	1	0	0	1	11	1	1	1	1	0	0	1	11	2
0	1	1	1	0	1	10	1	0	1	1	1	0	1	10	2
1	0	1	0	0	0	12	1	1	0	1	0	0	0	12	2

5. Positioning and motion control system of the mobile platform

The movement of the platform is carried out on the route specified by rectilinear orthogonal segments that run on the marking lines. The end coordinates of these segments correspond to the points of beginning and end of the movement, as well as intermediate points on crossroads, where change the direction of movement of the motionless platform. For each segment, the change of speed is carried out in accordance with the graph shown in Figure 7.

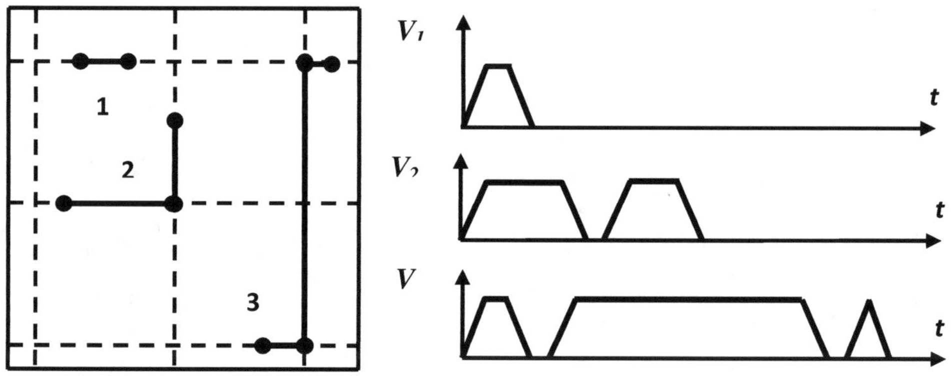

Figfure 7. The variation of speed on route segments.

The zone of acceleration, deceleration zone and zone of movement with nominal speed are allocated. The change in speed is carried out by changing the frequency of the pulses that are fed to the drivers of the drive stepper motors from the shaper CPSSM. The input signals for CPSSM are the signals of the direction of rotation of the stepper motor D and the synchropulses C of the variable frequency. The formation of signals C and D is carried out by the program of the microprocontroller MPU, which implements a mathematical model of platform motion based on the description of the route. This provides for uniformly accelerated and uniformly decelerated motion of the platform at the beginning and end of the segment. This is necessary to eliminate skipping of the stepper motor steps and slipping of the platform wheels. In the middle part of a segment uniform movement is carried out with speed which is defined by nominal speed of rotation of a shaft of the stepper motor reducer (0.25 turns / sec for the 28BYJ-28 engine). With a wheel diameter of 38 mm, this provides a nominal speed of 3 cm / sec. With the selected value of acceleration A = 3 cm / sec^2, the acceleration and deceleration time is 1 sec. With a reduction factor of 1/64, the nominal speed of the stepper motor shaft rotation is 16 turns / sec (1024 steps / sec for half-step mode). Uniform acceleration (deceleration) occurs over 512 steps and an decrement (increment) of the period of pulses is equal 3.8 ms. The distance L_a passed by the platform during acceleration (deceleration) is 1.5 cm. If the length of the segment $L>2\times L_a$, the time T_n of movement at nominal speed V_n is

$$T_n= (L-2\times L_a)/V_n \quad (1)$$

The total time T_s of movement between the start and end points of the segment will be equal

$$T_s= V_n/A+L/V_n \quad (2)$$

At the length of the segment $L\leq 2\times L_a$, the acceleration (deceleration) time is

$$T_a= (L/A)^{1/2} \quad (3)$$

The achieved maximum speed V_m will be equal

$$V_m= A\times T_a \quad (4)$$

The total number of steps K of the engine 28BYJ-28 (in half-step mode) for the passage of a segment of length L (in centimeters) is

$$K=L/0.0029 \quad (5)$$

The obtained ratios are necessary to determine the frequency and number of synchropulses in the formation of control pulses for stepper motors. They also provide the determination of the time of passage of the route crossroads (without stopping), which is necessary to synchronize the routes of the two platforms in order to eliminate collisions. The issues of planning and synchronization of routes of several platforms on the basis of the approaches presented in [8] are considered in more detail in a separate work. Determining the required number of engine steps to pass a given path does not provide accurate positioning of the platform due to wheels slippage. Therefore, the system provides the clarification of the position of the platform through the use of calibration markers on the marked surface, which are detected by the sensor S9. The signals of this sensor are processed by the MPU to refine and redefine the required number of steps.

Another problem solved by the platform motion control system is the exact positioning of the platform at the crossroad to change the direction of movement. To do this, for the direction of movement indicated by the arrow, optosensors S5-S9 are used. Sensor S9 captures the crossroad marker (with size 5×5 mm) and sensors S5-S8 provide accurate positioning of the platform at the intersection to change the direction of movement by 90 degrees for a motionless platform. When moving in the perpendicular direction, the signals of sensors S1-S4 and S5-S8 change places using sensors signal switch (SSS).

6. The use of the experimental model of a mobile platform in education process of CAD department

The planned use of the developed mobile platform in the process of diploma and course design by students of CAD Department involves solving the following tasks:

- Study and improvement of the structure and functional elements of the mobile platform.
- Research of dynamic characteristics of the mobile platform.
- Research of the tracking system for observance of a trajectory of movement on the set route.
- Research of the positioning system of the mobile platform and control of its movement.
- Planning and synchronization of routes for joint movement of platforms on the route network.
- Design and manufacture of platform components using 3-D printers and CNC machines.

Conclusion

In this work the following scientific and practical results are received:

- The kinematic scheme of the platform, which provides movement on orthogonal routes and tracking of the set trajectory of movement, is developed and realized.
- The structure of the digital control system of the mobile platform has been developed.
- The control subsystem of change the direction of movement and tracking of the trajectory, which set by lines of a marking, is developed and realized.
- The control subsystem of platform positioning and its movement with the calculated speed is developed and realized.

References

[1] Niku S B 2011 *An introduction to robotics: analysis, control, application* (New York: John Wiley & Sons)

[2] Shaha R Ozcelik S and Challoob R 2012 Design of a Highly Maneuverable Mobile Robot *Procedia Computer Science* **12** 170-5

[3] Novischi D M and Florea A M 2016 Ant Intelligent Robot: A Versatile and Low Cost Miniature Mobile Robot Platform for Swarm Robotics Research and Education *9th EAI International Conference on Bio-inspired Information and Communications Technologies (BIONETICS)*

[4] Kulpa M and Kolesnyk K 2016 The role of sensors in the electric drive wheel: Ways of enhancing informatory *functions XII International Conference on Perspective Technologies and Methods in MEMS Design (MEMSTECH)* 101-3 doi: 10.1109/MEMSTECH.2016.7507528

[5] Zdobytskyi A Lobur M Iwaniec M and Breznitskyi V 2019 Optimization of the Structural Characteristics of the Robotic System Holder *IEEE 15th International Conference on the Experience of Designing and Application of CAD Systems (CADSM)* 1-4 doi: 10.1109/CADSM.2019.8779311

[6] Denysyuk P Teslyuk V Kernytskyy A and Melnyk M 2019 A * Modification for Mobile Robotic Systems *IEEE XVth International Conference on the Perspective Technologies and Methods in MEMS Design (MEMSTECH)* 1-7 doi: 10.1109/MEMSTECH.2019.8817383

[7] Couceiro M S Figueiredo C M Luz J M A Ferreira N M and Rocha R P 2011 A low-cost educational platform for swarm robotics *International Journal of Robots, Education and Art*

[8] Mazur V 2013 Synchronization of City Bus Routes *Proc. Int. Conf. The Experience of Designing and Application of CAD Systems in Microelectronics* 249

IOP Conf. Series: Materials Science and Engineering 1016 (2021) 012012 doi:10.1088/1757-899X/1016/1/012012

A Unit for Inflating and Maintaining a Constant Pressure in the Tires During the Movement of a Slow-Moving Vehicle

KJ Waluś* and M Kozłowski

Poznań University of Technology, Piotrowo 3, 60-965 Poznań, POLAND

*Email: konrad.walus@put.poznan.pl

Abstract. In land transport, low-speed and special vehicles move over various terrain. From paved surfaces, through gravel and forest roads to boggy and muddy terrain, and the ability to overcome road obstacles depends, among others, on tire pressure. The obtained traffic parameters also depend on the environmental conditions, the condition and type of surface, the load, and the tire-surface cooperation area. The pressure in the tires of a moving vehicle has a significant impact on steering, driving stability, fuel consumption (ecology) and safety. Therefore, these vehicles are equipped with tire pressure systems. The article presents the conceptual design of the pumping unit and the constant pressure maintenance in the wheels of a slow-moving vehicle while driving. The presented unit is based on a hose pump located inside the wheel rim and a control system. For the designed structure, the values of forces and the time of inflating the tire to the given pressure were determined and subjected to FEM analysis.

Introduction

The environmental protection directives established by the Community have introduced standards for the emission of harmful compounds in the exhaust gases of motor vehicles. The European exhaust emission standard applies to new vehicles sold within the European Union [1]. The standards and norms were described in a series of directives that gradually increased their restrictiveness from the first standard called Euro 1, which came into force in 1993, to Euro 6 in 2014 [2-8] and the planned Euro 7 standard (currently in the social pre-consultation phase). The air resistance, the vehicle load [9-11] and the friction forces between the tires and the road surface have a significant impact on the increase in fuel consumption by vehicles [12-14]. Depending on the load of the vehicle, manufacturers recommend the appropriate pressure in the tires, which reduces the tire wear and affects the fuel combustion process. Maintaining the recommended pressure has an impact on the economy and ecology of the vehicle's movement. On November 1, 2014, the EU directive came into force, stating that all new vehicles must be factory-fitted with a pressure control system called TPMS (Tire Pressure Monitoring System).

The TPMS systems have been divided into two basic types: the indirect system and the direct system. The indirect system allows the tire pressure to be controlled via the ABS system. The sensors monitor the rotational speed of individual wheels, and the system calculates the rotation of a given wheel. The pressure drop in the tire reduces the rolling radius of the wheel, which causes the wheel to rotate more. The on-board computer compares the received data with the reference values and sends information to the dashboard when the set amount of air is lost. The system does not require additional equipment and special service procedures when replacing the tires. The direct TPMS system is equipped with pressure sensors, and the on-board computer displays the measured values in real time, or informs about too low pressure depending on the vehicle model and the installed equipment.

The movement of vehicles, apart from the driver's abilities, depends on the transferred forces and moments acting in the area of tire-road interaction [15-18]. The vehicle motion parameters also depend on the adhesion coefficient, which is determined by the features of the road surface [19-21], the tires [22-24], the structure of the vehicle's drive and suspension [25, 26] and the features of the environment [27, 28]. Apart from external factors, it is also the vehicle design changes that have an impact [29-31].

Moving on marshy and soft terrain, the possibility of transferring forces and moments at the contact of the tire with the ground is used at reduced tire pressure. Lowering the pressure increases the contact the surface area and reduces the unit pressures [31]. Suitable tires are used in off-road, agricultural, slow-moving and special vehicles to more effectively overcome unfavorable terrain. Driving such a vehicle on an uneven terrain, overcoming various obstacles, often has a negative effect on the tires and the suspension [29, 32, 33]. By controlling the tire pressure, we influence the traffic safety, the traction, the vehicle steerability, the rolling resistance and the fuel consumption.

The tire pressure is easily adjustable and controllable and plays an important role in reducing skid on unpaved roads. This aspect greatly affects the fuel consumption. Damanauskas et al. in 2015 show that reducing tire pressure reduces drive wheel slip and fuel consumption, while increasing work efficiency, and the occurrence of skid in the range of 7 to 15% can be considered normal during field work [34]. In the case of wheel slip control in conjunction with changes in the tire pressure, Janulevičius and Damanauskas in 2015 predict the possibility of minimizing the fuel consumption [35].

Correctly selected pressure enables better contact with the surface, limiting the slip phenomenon. This translates into more effective use of the available power of the machine. Less energy is wasted on movement, thus saving fuel and time for work (ecology + economy). However, reduced tire pressure increases the radial deflection and negatively affects the heat balance of the tire composite layers during motion. Similarly, too low tire pressure and reduced tension value of composite drive belts cause their premature wear [36 - 41].

The influence of tire pressure on skidding

Tire manufacturers usually list the permissible pressure of the tires they sell. This value is not optimal for the work to be performed. For example, at the permissible pressure, e.g. 0.20 MPa, works can be most effective at the pressure of 0.12 MPa. This is due to the increased contact area and greater use of the tread pattern. Tests carried out by Michelin show a strong correlation between the slip of the wheels and the tire pressure. It has been shown that lower pressure causes lower wheel slip value during field works. It was noticed that for the field works performed at the speed of 10 km / h, the pressure of 0.25 MPa caused a slip of 38%, while the pressure of 0.10 MPa - 17%. It can be seen that the difference of 0.15 MPa allows for reducing the wheel slip value by as much as 21%.

The currently used central pressure regulation systems in wheels are characterized by high complexity of the structure. The key element is the use of a central pumping unit, which requires the use of pneumatic lines and connecting them to a moving valve located on the rim. Today there are two constructional solutions enabling the air supply to the wheels. In structures equipped with portal bridges, the possibility of making a "through" axis is used. The medium is supplied by feeding the cable through the central axis of the hub and connecting the cable to a dedicated adapter. Forming a sliding pair with the lead-out, the end of the tubing connects the end of the tubing to the rim valve. This prevents the pneumatic hoses from twisting. In other cases, the medium is supplied through a connection mounted above the wheel and, similarly to portal bridges, an adapter is used to connect the pipes to the valve on the wheel rim.

There are also methods for delivering air to the tires through the special construction of the rim. It then has a hollow frame through which the medium is delivered to the inside of the tire. In this case, it is also necessary to use a special adapter in the centre of the wheel.

Table 1. Comparison of the features of compressed air supply to the inside of the tire

	disadvantages	advantages
power cord routed through the reduction gear pinion	- necessity to use a rotating head	- low cost of production - low operating costs
the power cord routed outside the vehicle	- frequent mechanical damage to the cord - necessity to use a rotating head	- possibility of using a larger diameter hose (shorter pumping time)
power cord routed through the wheel drive shaft	- system leakage problems - complicated adaptation of the drive system (increased costs) - weakening of the drive system - necessity to use a rotating head	- avoiding the breakage of the guide cord
supplying of the compressed air through the hollow rim shoulder	- significant cost of a specialized reinforced rim with a hollow shoulder - complicated adaptation of the drive system (increased costs) - weakening of the drive system - necessity to use a rotating head	- avoiding the breakage of the guide cord - easy sealing of the system by minimizing the number of pipe connections

Another example is the solution proposed by Mercedes-Benz. The system is equipped with a central compressor and a pneumatic installation system. The difference to other solutions is the use of pressure tanks as compressed air accumulators. These elements are embedded in the wheel arches of the vehicle. In the system, the compressor is responsible for maintaining high pressure in the tanks. This solution allows for raising the pressure in the tires from 0.05 MPa to 0.18 MPa in about 20 seconds.

Most modern constructions are based on a common concept of equipping the vehicle with a central pumping unit and supplying the air to the wheels while the vehicle is moving. This affects the complexity of these solutions, which directly translates into the system costs and difficulties in adapting to the already existing vehicles. These systems are prone to a number of failures, and a failure of the central pumping unit practically means the loss of the possibility of pressure adjustment. The mentioned wheel pumping systems have a number of drawbacks, such as [own elaboration]:

- the complicated system of pneumatic connections,
- the need to use adapters to obtain a rotary pneumatic connection,
- problems with the occurrence of mechanical damage,
- a large number of connections favours the formation of leaks.
- the weakening of the drive system (in the case of using solutions consisting in supplying the medium through the drive shaft),
- a failure of the central unit paralyzes the operation of the system,
- low adaptation possibilities,
- high system costs.

The ability to remotely adjust the pressure in the tires due to the tire-surface cooperation determines the need to design a new construction of the tire inflation system for the slow-moving vehicle while driving. The ability to quickly and smoothly adjust the pressure for such machines is a desirable feature that can affect the comfort, cost and culture of operation of such machines. The paper presents a conceptual design of a pumping unit and a constant pressure of pneumatic wheels during the movement of a slow-moving vehicle

Conceptual design of the inflating unit

The use of portal bridges in vehicles enables them to overcome higher obstacles and reduces the probability of damage. More and more often, portal bridges are also used to make it easier to equip a vehicle with a central tire inflation system. An alternative solution may be to use a peristaltic pump.

IOP Conf. Series: Materials Science and Engineering 1016 (2021) 012012 doi:10.1088/1757-899X/1016/1/012012

The main feature of peristaltic pumps is their tightness. The pumped medium does not come into direct contact with the mechanical parts of the pump. This results in a significant reduction in the possibility of contaminants getting inside the tire. This pump is also characterized by a simple structure limiting the elements present in it. The pumping of substances is possible using one pipe, the ends of which are connectors, and the pipe itself is the working section of the pump. The drawback of the presented concept is the impossibility of using the structure for steering axles, as well as the need to supervise and plan the replacement of working cables due to their wear and degradation. Changing of the orientation of the wheel in relation to the driving axle may result in the damage and destruction of the inflating unit.

The construction and design assumptions

The assumption of the structure is to maintain its full functionality, to inflate the wheels while the vehicle is in motion, and to minimize the modifications to the existing machine structure. The conceptual design assumes interference only with the vehicle wheels. This involves the necessity to make an additional hole in the rim and to equip it with an additional ring with an elastic element. In order to implement the design assumptions, it was decided to install a mechanism based on the construction of peristaltic pumps inside the wheels of the vehicle. A view of the pneumatic wheel inflating unit while driving and maintaining a constant pressure is shown in Figure 1.

The device carries out the process of inflating the wheels without an additional pump that would require an external drive. The pneumatic wheel inflating unit while driving and maintaining a constant pressure is intended for installation in non-driven and non-steered wheels. In such a solution, the wheel is made of a pneumatic tire 1, mounted on a rim 2 adapted for mounting the axle 6, a flexible pneumatic channel 4 and pneumatic valves: a suction valve 14, a pumping valve 15 and a pressure control valve 16. A cross 5 is mounted on the stationary axis 6. A flexible pneumatic duct 4 is permanently attached to the rim 2 from the inside. Flexible pneumatic duct 4 is equipped with a suction valve 14 and a pumping valve 15 fixedly attached to the rim. The pressure for inflating the wheel is generated by the rolling of the rollers 3 on the flexible pneumatic duct 4. The air in the flexible pneumatic duct 4 is forced only while the vehicle is moving. The continuous supply of air to the pneumatic tire could contribute to its damage, therefore a pressure control valve 16 was used.
This approach eliminates the drawback of the central unit by equipping each wheel with a separate air delivery unit. The independence of the inflating units in the event of failure of one of the systems enables further pressure adjustment of the other wheels and the maintenance of the overall efficiency of the vehicle. The encapsulation of the pressed elements inside the rim reduces the possibility of damage to the elements responsible for transmitting the air to the inside of the tire, e.g. by their accidental breaking. To simulate a system, it is necessary to perform preliminary calculations and determine the forces acting on the structure. The calculations assumed that air is a perfect gas and its parameters correspond to the characteristics of the isentropic transformation.

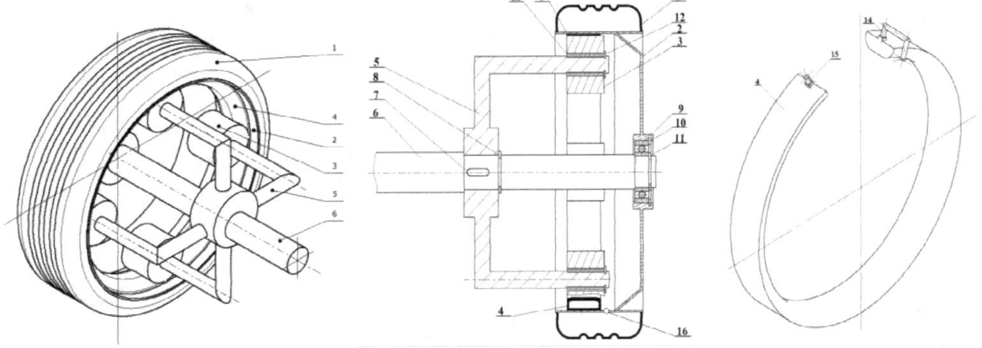

Figure 1. Conceptual design of the pneumatic wheel inflating unit while driving and maintaining a constant pressure (own study): 1 pneumatic tire, 2 rim, 3 rollers, 4 flexible pneumatic duct, 5 cross, 6 assembly axis, 7 key connection, 8 spring snap ring, 9 bearing , 10 internal snap ring, 11 external snap ring, 12 snap ring, 13 slide bearing, 14 suction valve, 15 pumping valve, 16 pressure control valve.

The requirement for the proper functioning of the unit is the minimum pressure difference necessary to open the one-way inflating valve 15. For the calculations, it was assumed that the required pressure difference was 0.001 MPa. The force acting on the shoulder of the cross will be equal to the pressure acting on the surfaces of the elastic element. The assumed cross-section of the elastic element is:

$$S = \frac{\pi d^2}{8} + h(l-d) = \frac{\pi 60^2}{8} + 30(110-60) = 2913{,}72 mm^2$$

Assuming that air is the ideal gas and the assumed pressure to open the valve is, e.g. 0.201 MPa, the force of impact on the arm will be:

$$F = P \cdot S = 0{,}201 \cdot 2913{,}72 = 585{,}66 N$$

where:
F - force of impact on the arm,
P - pressure to be achieved on the pressed side,
S - area affected by the pressure.

The value of the force acting on the arm depends on the set tire pressure. For the selected range of pressures, the values of the forces necessary to overcome the resistance of the rollers movement along the flexible pneumatic channel were presented (Table 2).

Table 2. Table of forces acting on the structures (own study)

Target pressure [MPa]	Minimum compression pressure [MPa]	Calculated force acting on the frame [N]
0.150	151	439.97
0.175	0.176	512.81
0.200	0.201	585.66
0.225	0.226	658.49
0.250	0.251	731.34

The second important parameter is the pumping time, which depends on the size of the tire used and the speed of the vehicle. For example, for a tire of the size R26 480/80, the projected necessary vehicle speed to inflate the wheel from 0.15 MPa to 0.25 MPa to 1 minute is approximately 17 km / h. This speed is the value achievable by a slow-moving vehicles, since the limit speed is 25 km / h.

Numerical analysis - FEM

The designed structure of the pneumatic wheel inflating unit while driving and maintaining a constant pressure is shown in Figure 2. While the vehicle is moving, the wheels rotate causing the apparent movement of the cross with rollers on the elastic element. The structure must be characterized by strictly defined geometrical relations to enable the inflating process.

The location of the structure in the vehicle and the function it performs are strictly determined by the strength criterion. The cross with the rollers is the most heavily loaded. Under normal conditions, when working in the setting that maintains the pressure of 0.25 MPa in the tires, the maximum value of the force acting on the rollers of the cross is the 730N wheel. The structure is equipped with four rollers and each of them is subjected to variable loads during movement. During the analysis, for safety reasons, the same force value of 1000 N was assumed for each of the rolls. The areas of higher stress values are the support frame and the roller seats. The results show negligible stresses in the main plate of the cross. In order to investigate specific important points in the structure in more detail, a probe was used to read the stress values. The system of forces and the grid structure for numerical calculations are shown in Figure 3.

IOP Conf. Series: Materials Science and Engineering 1016 (2021) 012012 doi:10.1088/1757-899X/1016/1/012012

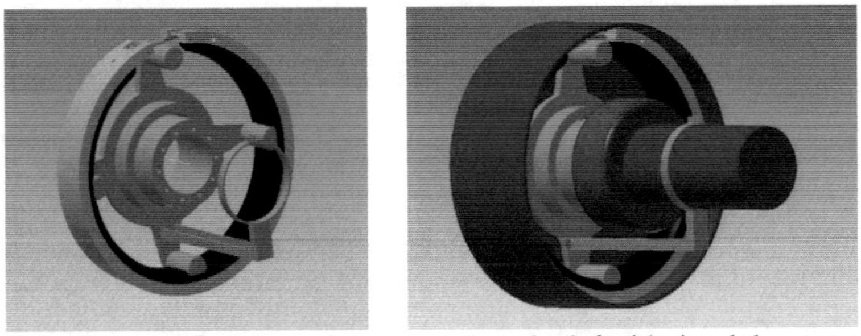

Figure 2. View of the 3D model of the entire system / on the left side / and the mutual relations of the designed structure in the system of basic vehicle elements / on the right side / (own study).

For the adopted boundary conditions, the maximum stress values were 219.1 MPa. The maximum stress values occur on the support and do not exceed the yield point of the material accepted. The greatest displacements are 2.41 mm and occur at the ends of the arms of the cross. The displacement analysis performed shows that the entire structure retains the required stiffness, as shown in Figure 4. The values of safety factors were determined for the entire structure. The designed cantilever has a factor above 1.5, and the main structure plate from 7 to 15.

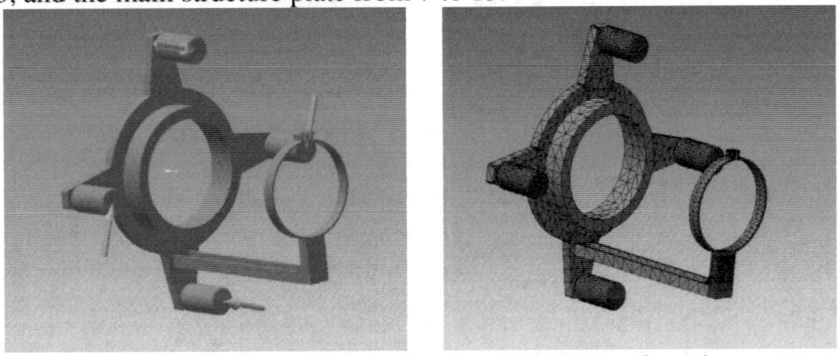

Figure 3. Presentation of the application of forces and mesh structure in the FEM simulation (own study).

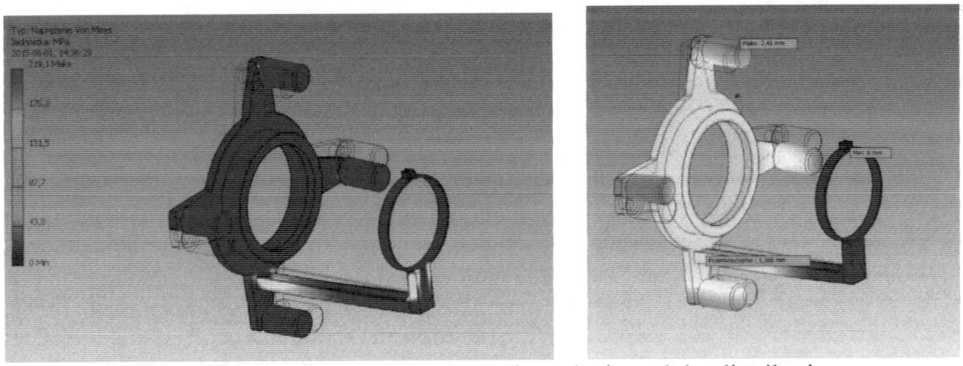

Figure 4. The results of the strength analysis and the distribution of displacements during the load on the structure (own study).

In addition to the mechanical structure, it is necessary to introduce a tire pressure control system. By changing the settings on the control panel, the pressure adjustement valve is controlled. The operation of the control system should be divided into the transmitting and receiving part. The transmitting system should be based on a microcontroller that allowing for controlling the settings of the pressure adjustement valve and collecting the information about the current pressure in the tires. The receiving system is present in all wheels on which the inflating unit is mounted. It consists of the receiver signal, and executive part of the control relay is a solenoid valve.

Conclusions

The wheel inflating systems are playing an increasingly important role in vehicles. A number of benefits result in great interest in adapting such systems to military and civilian applications.

The developed structure of the system is an innovative solution, not present before in wheel inflating systems. The possibility of easy adaptation and the lack of integration in the basic structure of cooperating machines makes the developed system unprecedentedly versatile. Equipping the machine with the additional feature of tire pressure adjustement could therefore take place in a very short time, and the process itself would resemble wheel replacement. Other features, such as the lack of a central inflating unit or the lack of elements protruding beyond the outline of the vehicle, translate into a reduced failure rate of the system.

The designed system has great opportunities for further development of the construction. Equipping the cross with appropriate gears and drive will result in the possibility of its operation while the vehicle is not in motion. Such a functionality and development of the control program may result in active monitoring of the pressure in the wheels.

In order to thoroughly analyze the system, it would be necessary to make a prototype and subject it to a series of tests. Such tests would show the actual parameters achieved by the system and would provide data on the existence of possible risks related to the strength of individual components. Such collected complete data would allow for the improvement of the construction and for its adaptation to proper functioning in real conditions.

References

[1] Waluś K J Warguła Ł Krawiec P and Adamiec J M 2018 Legal regulations of restrictions of air pollution made by non-road mobile machinery – the case study for Europe: A review *Environmental Science and Pollution Research* **25** 3243-59 https://doi.org/101007/s11356-017-0847-8

[2] Council Directive 93/12/EEC of 23 March 1993 relating to the sulphur content of certain liquid fuels

[3] DIRECTIVE 94/12/EC OF THE EUROPEAN PARLIAMENT AND THE COUNCIL of 23 March 1994 relating to measures to be taken against air pollution by emissions from motor vehicles and amending Directive 70/220/EEC

[4] European Parliament and Council Directive 94/63/EC of 20 December 1994 on the control of volatile organic compound (VOC) emissions resulting from the storage of petrol and its distribution from terminals to service stations

[5] Directive 98/70/EC of the European Parliament and of the Council of 13 October 1998 relating to the quality of petrol and diesel fuels and amending Council Directive 93/12/EEC

[6] COMMISSION REGULATION (EC) No 692/2008 of 18 July 2008 implementing and amending Regulation (EC) No 715/2007 of the European Parliament and of the Council on type-approval of motor vehicles with respect to emissions from light passenger and commercial vehicles (Euro 5 and Euro 6) and on access to vehicle repair and maintenance information

[7] REGULATION (EC) No 661/2009 OF THE EUROPEAN PARLIAMENT AND OF THE COUNCIL of 13 July 2009 concerning type-approval requirements for the general safety of motor vehicles their trailers and systems components and separate technical units intended therefor

[8] COMMISSION REGULATION (EU) No 214/2014 of 25 February 2014 amending Annexes II IV XI XII and XVIII to Directive 2007/46/EC of the European Parliament and of the Council establishing a framework for the approval of motor vehicles and their trailers and of systems components and separate technical units intended for such vehicles

[9] Kukla M Wieczorek B Warguła Ł and Berdychowski M 2019 An analytical model of the demand for propulsion torque during manual wheelchair propelling Disability and Rehabilitation *Assistive Technology* 1-8

[10] Wieczorek B and Kukla M 2019 Effects of the performance parameters of a wheelchair on the changes in the position of the centre of gravity of the human body in dynamic condition *PloS one* **14(12)** e0226013

[11] Wieczorek B and Kukla M 2020 Biomechanical Relationships Between Manual Wheelchair Steering and the Position of the Human Body's Center of Gravity *Journal of Biomechanical Engineering* **142(8)**

[12] Wieczorek B and Warguła Ł 2019 Problems of dynamometer construction for wheelchairs and simulation of push motion *MATEC Web of Conferences* **254** 01006 https://doi.org/101051/matecconf/201925401006

[13] Pałasz B Waluś K J and Warguła Ł 2019 The determination of the rolling resistance coefficient of a passenger vehicle with the use of selected road tests methods *MATEC Web of Conferences* **254** 04006 https://doi.org/101051/matecconf/201925404006

[14] Pałasz B Waluś K J and Warguła Ł 2019 The determination of the rolling resistance coefficient of a passenger vehicle with the use of roller test bench method *MATEC Web of Conferences* **254** 04007 https://doi.org/101051/matecconf/201925404007

[15] Olszewski Z and Waluś K J 2012 Effects of Atmospheric Conditions on The Mechanical Properties of Tires *Applied Mechanics and Materials* **232** 14-18

[16] Waluś K J and Olszewski Z 2011 Analysis of Tire-road contact under Winter conditions *Lecture Notes in Engineering and Computer Science* **2192** *World Congress on Engineering* **3** 2381-4

[17] Waluś K J 2017 Driver's Strategy and Braking Distance in Winter *Transport Means Proceedings of the 21st International Scientific Conference* 505-9

[18] Farhadi P Golmohammadi A Sharifi A and Shahgholi G 2019 The CIGR e-journal Prediction of the tractor tire contact area contact volume and rolling resistance using regression model and artificial neural network *Agricultural Engineering International* **21**(3) 26-38

[19] Sawicki P Waluś K J and Warguła Ł 2018 The comparative analysis of the rolling resistance coefficients depending on the type of surface – experimental research *Transport Means Proceedings of the 22nd International Scientific Conference* 434-41

[20] Kędziora K and Waluś K J 2013 Experimental and simulation research on car acceleration and braking on snow-covered roads *Activities of Transport Telematics 13th International Conference on Transport Systems Telematics TST 2013: selected papers* ed Jerzy Mikulski: Springer Berlin Heidelberg Communications in Computer and Information Science **395** 433-40

[21] Waluś K J Polasik J Mielniczuk J and Warguła Ł 2019 Experimental tests of vehicle body acceleration at selected railway crossing *MATEC Web of Conferences* **254** 04002 https://doi.org/101051/matecconf/20192540400

[22] Warguła Ł Wieczorek B and Kukla M 2019 The determination of the rolling resistance coefficient of objects equipped with the wheels and suspension system–results of preliminary tests *MATEC Web of Conferences* **254** 01005

[23] Warguła Ł Kukla M Wieczorek B 2020 IOP The impact of wheelchairs driving support systems on the rolling resistance coefficient *Conf Series: Materials Science and Engineering* **776** 012076 https://doi.org/101088/1757-899X/776/1/012076

[24] Kulikowski K and Szpica D 2014 Eksploatacjai Determination of directional stiffnesses of vehicles' tires under a static load operation *Niezawodnosc – Maintenance and Reliability* **16** (1) 66–72

[25] Wieczorek B Warguła Ł and Rybarczyk D 2020 Impact of a Hybrid Assisted Wheelchair Propulsion System on Motion Kinematics during Climbing up a Slope *Applied Sciences* **10** 1025

[26] Waluś K J Polasik J Krawiec P and Warguła Ł 2017 The research on the intensity of sports car acceleration with the use of launch control system *Transport Means Proceedings of the 21st International Scientific Conference* 564-8

[27] Martin D P and Schaefer G F 1996 Tire-Road Friction in Winter Conditions for Accident Reconstruction *SAE Technical Paper Series* 960657
[28] Navin F Macnabb M and Nicolletti C 1996 Vehicle Traction Experiments on Snow and Ice *SAE Technical Paper Series* 960652
[29] Dižo J Blatnický M Sága M Harušinec J Gerlici J and Legutko S 2020 Development of a New System for Attaching the Wheels of the Front Axle in the Cross-Country *Vehicle Symmetry* **12** 1156
[30] Waluś K J Warguła Ł Krawiec P and Adamiec J M 2017 The impact of the modernization of the injection-ignition system on the parameters of motion of the motorcycle *Procedia Engineering* **177** 393-8
[31] Polasik J Waluś K J and Warguła Ł 2017 Experimental studies of the size contact area of a summer tire as a function of pressure and the load *Procedia Engineering* **177** 347-51 https://doi.org/101016/jproeng201702203
[32] Polasik J and Waluś K J 2016 Analysis of the force during overcoming the roadblock – the preliminary experimental test *Transport Problems* **11**(1) 113-20 https://doi.org/1020858/tp201611111 ISSN 1896-0596
[33] Waluś K J Polasik J Mielniczuk J and Warguła Ł 2019 Experimental tests of vehicle body accelerations at selected road and rail crossings *MATEC Web of Conferences* **254** 04002-1-04002-7 https://doi.org/101051/matecconf/201925404002
[34] Damanauskas V Janulevičius A and Pupinis G 2015 Influence of extra weight and tire pressure on fuel consumption at normal tractor slippage *Journal of Agricultural Science* **7**(2) 55-67
[35] Janulevičius A and Damanauskas V 2015 How to select air pressures in the tires of MFWD (mechanical front-wheel drive) tractor to minimize fuel consumption for the case of reasonable wheel slip *Energy* **90** 691-700
[36] Czarnecka-Komorowska D and Tomczyk T 2008 Influence of montmorillonite (MMT) on the mechanical properties in recycled rubber *Arch Mech Technol Autom* **28**(1) 145-52
[37] Czarnecka-Komorowska D and Wiszumirska K 2020 Sustainability design of plastic packaging for the Circular Economy *Polimery* **65** 8-17
[38] Czarnecka-Komorowska D Wiszumirska K and Garbacz T 2018 Films LDPE/LLDPE made from post-consumer plastic: processing structure mechanical properties *Adv Sci Technol Res J* **12** 134-42
[39] Krawiec P Waluś K J Warguła Ł and Adamiec J 2018 Wear evaluation of elements of V-belt transmission with the application of optical microscope *MATEC Web of Conferences* **157** 01009-1-8 https://doi.org/101051/matecconf/201815701009
[40] Krawiec P Waluś K J Warguła Ł and Adamiec J 2019 Wear evaluation study of the multiple grooved pulleys with optical method *MATEC Web of Conferences* **254** 01004-1-8 https://doi.org/101051/matecconf/201925401004
[41] Krawiec P Warguła Ł Małozięć D Kaczmarzyk P Dziechciarz A and Czarnecka-Komorowska D 2020 The Toxicological Testing and Thermal Decomposition of Drive and Transport Belts Made of Thermoplastic Multilayer Polymer Materials *Polymers* **12** 2232

IOP Conf. Series: Materials Science and Engineering 1016 (2021) 012013 doi:10.1088/1757-899X/1016/1/012013

Diagnostics of the Thermal Condition of the Cable Gear Used in the Drive of a Wood Chipper

P Krawiec, Ł Warguła and L Różański

Poznań University of Technology, Piotrowo 3, 60-965 Poznań, POLAND

*Email: piotr.krawiec@put.poznan.pl

Abstract. Early detection of the causes of malfunction of toothed gears, cable gears, and various types of drives enables the prevention of breakdowns and related downtime. The work concerns the possibility of using terminography in order to assess the thermal state of the V-belt of the tension transmission used in the drive of a wood chipper. This device is exposed to dynamic loads and one of the main elements transmitting the drive is a cable gear with a V-belt. The paper presents the use of thermography to evaluate the operation of a cable gear with a heat-welded V-belt. As a diagnostic characteristic, the dependence of temperature changes of the belt and pulleys on the variable load of the chipper over time was adopted. The variable load on the device results from the feeding time of the branches and their cross-section. The aim of the work was to investigate the influence of variable transmission load on the durability and life of the V-belt. It has been found that for the cable transmission it is desirable to ensure that the temperature of the belt around the circumference is similar and changes little.

1. Introduction

The use of V-belts in cable transmissions dates back to the beginning of the 20th century, however, in the last dozen or so years, the intensive development of industry contributed to the improvement of the existing, developed forms of geometric drive and transport belts. The cause of improper operation of the traction gears may be mistakes resulting from incorrect selection of geometrical features of the gear, errors occurring during the manufacture of wheels, incorrect assembly of the gear [1-3], incorrect geometry of details during spinning [4-6], premature destruction of parts [7-8], change in temperature of the gear [9-10] the last feature may cause the belt to elongate too quickly and, consequently, destroy it. The V-belts used in cable gears are most often made of composite materials with a composition based on plastics [11-14]. So far, the production of tension belts has traditionally used steel or polyamide cords constituting the load-bearing layer, rubber or caoutchouc to make the flexible layer, and a fabric-rubber composite for the protective layer. Due to their mechanical properties, V-belts are used in transport systems where other strings do not meet the functional requirements [15-17]. Examples of such applications are conveyors transporting products at high speeds, exposed to significant abrasive wear, in difficult environmental conditions [18, 19]. Thermography is used to diagnose kinematic nodes in rotary mowers [20] or to analyze the stability of the process of extrusion of microporous polyvinyl chloride (PVC). In the literature, you can find many examples of the evaluation of belt transmissions with flat, wedge and toothed belts where thermal imaging cameras were used [21-25]. The principle of operation of such solutions is based on the indirect measurement of the temperature of the analyzed object, while the emission power generated by the object is directly assessed [26-31].The emission of the object is converted, using the system's detection system, into an electrical impulse which is a means of informing about the thermal state of the object. The thermograph makes it possible to determine the temperature, while the thermometric characteristics are placed on the computer disk which enables the visualization of the thermograph by the camera.

During the temperature from the application of thermovision values, errors (error of the emissivity test of the object, error related to the error test) and calibration, errors related to the measurement test. The principle of carrying out thermal imaging measurements is described in [9].

The objective of the study was to assess the thermal condition of the cable gear used in the drive of a wood chipper. The knowledge of the temperature changes of V-belts is important when selecting both the type of belt and possible applications. The test results can be used in the process of selecting belts used in heavily loaded transmissions as well as for people involved in work safety in enterprises using machines and devices with cable transmissions.

2. Research methodology

2.1. Subject of research

A belt drive of a wood chipper [32-34] with two A 1500 V-belts operating in parallel and a cross-section of 13x8 was selected for the study. Due to the color and lack of gloss (the wheels were covered with rust coating), it was not necessary to paint them. The experiments used a thermal imaging camera called Termovision FLIR T620, which has a spatial resolution of 0.62 mrad, the accuracy of measurements for this camera is ± 2 ° C or ± 2% of the measured value.

2.2. Test stand and the course of examinations of the thermal state of the gear

The tests of the cable gear driving the wood chipper were carried out in December at an ambient temperature of -2^{about}C. The driving wheel was mounted on the motor shaft, and the driven wheel was the shaft drive on which the chipper knives were mounted. The gear worked as a reducer, the ratio was 1: 5. The gear load was applied by periodically feeding the branches to the chipper. The course of the research was as follows: after starting the station and idling for 5 minutes, the branches were supplied to the chipper. In the first stage, this work was performed by one operator. After filling the container, there was a short break for its replacement and then the material was supplied to the device by two operators. Due to the fact that the cross-section of the branches was different, it generated a variable load on the transmission. When branches with a diameter of more than 70 mm were given, the drive sometimes stopped and the temperature rose sharply. The following distances of the branches from the chipper were assumed (3, 9, 15 meters) [35]. The test stand for the assessment of the impact of load on the temperature of V-belts is shown in Figure 1.

Figure 1. Test stand to assess the impact of load on the temperature of V-belts.

3. Research Results

The result of the performed experimental tests are information on the values of the current temperature of the belts recorded on thermal images (thermograms) and their distribution illustrated on the profilograms. To facilitate the assessment, the red color is assigned to the active cable, while the green color is the passive cable. In order to carry out thermal analysis, a specialized computer program ThermaCAM Reporter 2000 Professional was used. Profile lines of the active and passive tendon were assessed (Fig. 2).

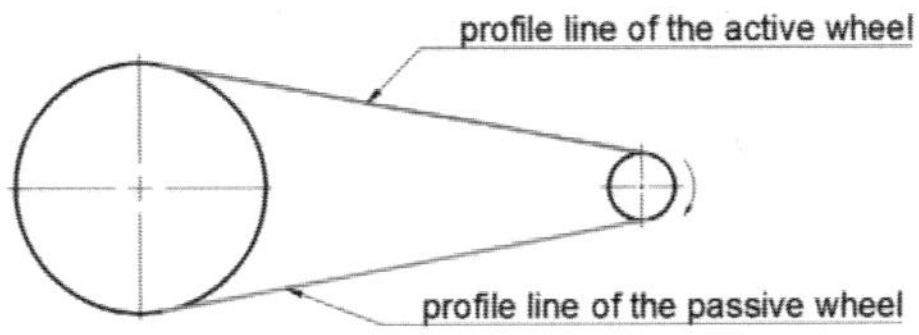

Figure 2. The designation adopted in the temperature measurement procedure of the tension gear belts with a V-belt.

The instantaneous temperature values during the gearbox operation without load are shown in Figure 3a, while the temperature distribution for these operating conditions is shown in Figure 3b. During this time, the camera recorded the absolute temperature difference as a value of 0 ° C between the temperature in the active (upper) cable and the temperature in the passive (lower) cable. The analysis of Figures 3a and 3b shows that the temperature difference of $\Delta T=0.0^{degrees}$ indicates the correct operation of the measuring apparatus.

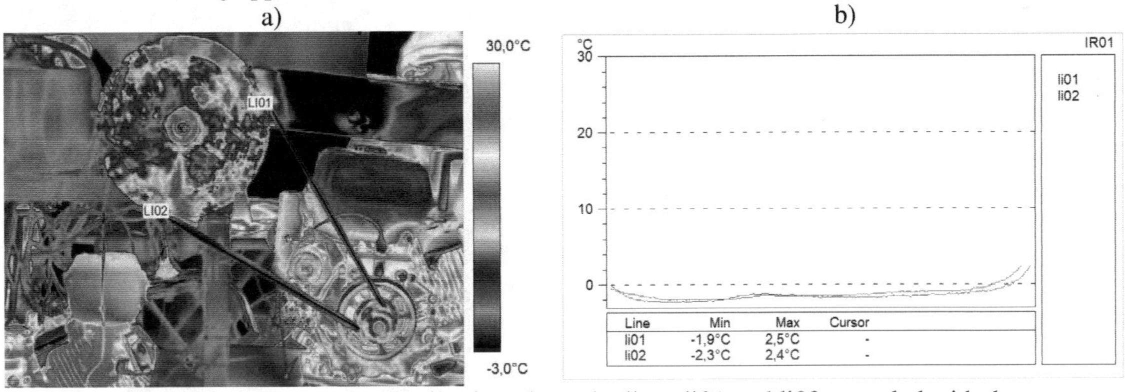

Figure 3. Temperature distribution along the lines li01 and li02 recorded with the transmission stationary, red - active cable, green - passive cable.

Figure 4 shows the temperature change on the tendon surface after 1.32 minutes, when the weight of the chips was 33 kg. The branches were fed to the device by one operator. The distance of the branches from the shredder was 3 meters. An increase in the mean temperature was observed in the active cable (38.8 °C) in the relaxation phase and in the inactive cable (33.7 °C) in the dropping phase from the driving wheel. The recorded temperature change of the belt was caused, on the one hand, by its significant stretching when starting cooperation with the driving pulley, and on the other hand, by its compression before leaving the cooperation (Fig. 4a).

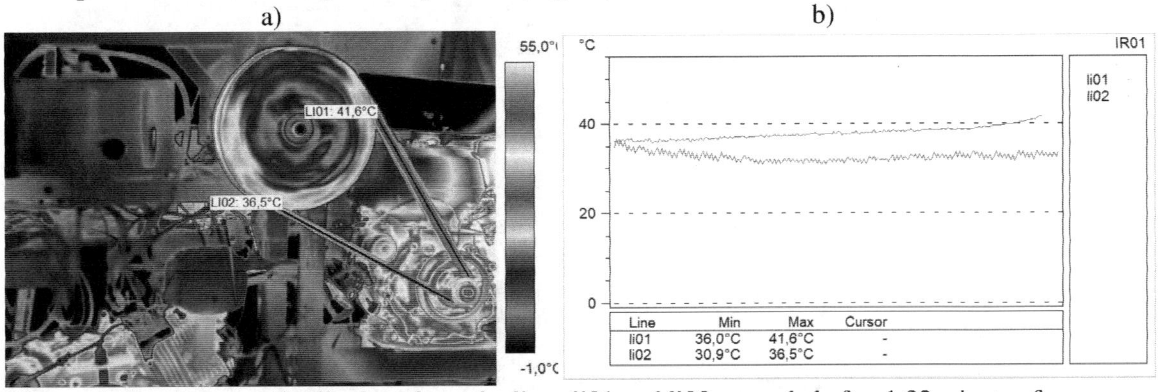

Figure 4. Temperature distribution along the lines li01 and li02 recorded after 1.32 minutes from the commencement of comminution, (one operator, distance between the branch and the chipper 3 meters) red - active cable, green - passive cable.

In subsequent experiments, Figure 5 shows the temperature distribution on the V-belt and the profilograms recorded during the work of two operators. The temperature was recorded after 15 minutes when the weight of the wood chips obtained was 208 kg. In this test, a significant increase in the average temperature was recorded both in the active cable (approx. 73.3 ºC) and the passive one (64.75 ºC), the difference in temperature was ΔT = 8.55 ^{degrees}C. The ambient temperature was then 0 ºC.

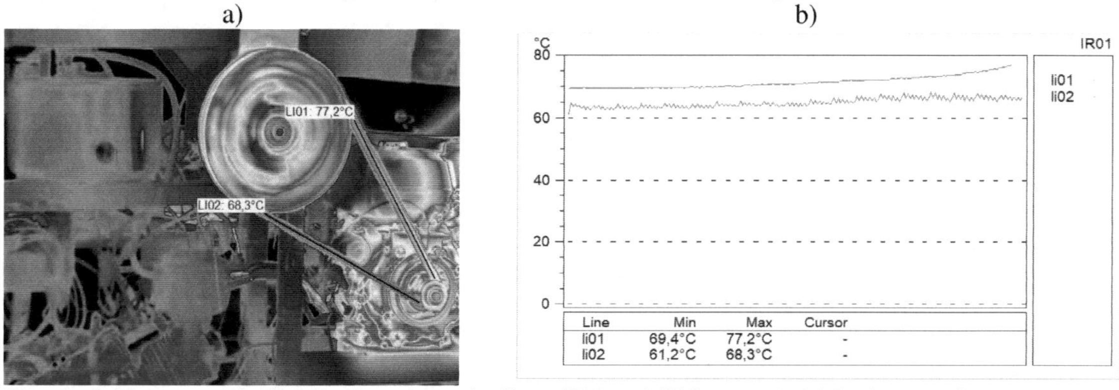

Figure 5. Temperature distribution along the lines li01 and li02 recorded 15 minutes after commencement of comminution, (one operator, distance of the branch from the chipper 3 meters) red - active cable, green - passive cable.

The next experiments were carried out 26 minutes after the commencement of grinding, then a total of 402 kg of wood chips was obtained, the results are included in the drawing 6. The analysis of the obtained belt temperature distributions and the performed profilograms indicate a gradual reduction and equilibrium of the temperature in the in the active (approx. 37.9 ºC) and passive (34.7 ºC) cable was observed. There were vibrations in the active band which additionally lowered the temperature of the belts.

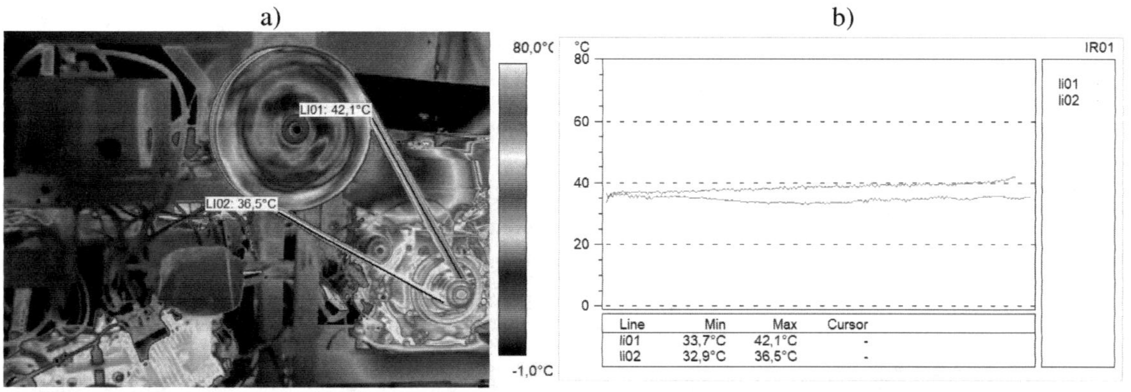

Figure 6. Temperature distribution along the lines li01 and li02 recorded 26 minutes after commencement of comminution, (one operator, the distance of the branch from the chipper 3 metres) red - active cable, green - passive cable.

In turn, Figure 7 shows the change in belt temperature and recorded profilograms recorded during the second stage of the tests at a distance of 9 metres, but branches with a diameter of about 100 mm were given. After producing the first 30 kg of chips, a very rapid increase in the average temperature was observed in the active cable (75.9 ° C) during entering and in the passive cable (67.6 ° C) when leaving the driving wheel. Then, with the adopted rules for conducting the experiment, an increase in the temperature difference between its values in the active and passive cables was observed, equal of ΔT = 8.3 °C. Such a high temperature of the belts was caused by the operation of the device during very unfavorable working conditions.

a) b)

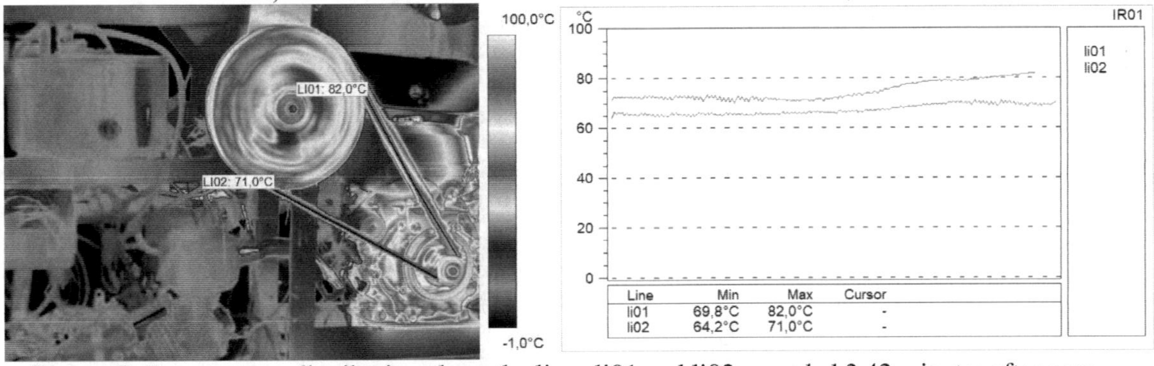

Figure 7. Temperature distribution along the lines li01 and li02 recorded 2,42 minutes after commencement of comminution, (one operator, the distance of the branch from the chipper 9 metres) red - active cable, green - passive cable.

After 26 minutes, further experiments were carried out using thermal imaging technology. After this time, 189.5 kg of wood chips were produced. The recorded values of instantaneous temperatures and their distribution on the circumference of the belt are included in Figures 8a-b. It was observed that the difference between the temperature in the active (77.6 ºC) and passive (70.6 ºC) tendons, equal to $\Delta T = 7$ ° C. It can be concluded that there is a direct correlation between the temperature of the wheels and the temperature of the V-belt.

a) b)

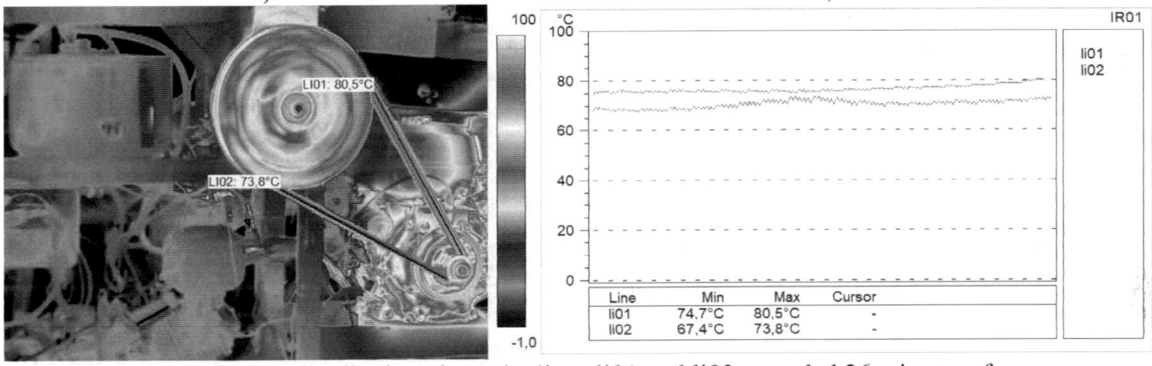

Figure 8. Temperature distribution along the lines li01 and li02 recorded 26 minutes after commencement of comminution, (one operator, the distance of the branch from the chipper 9 metres) red - active cable, green - passive cable.

a) b)

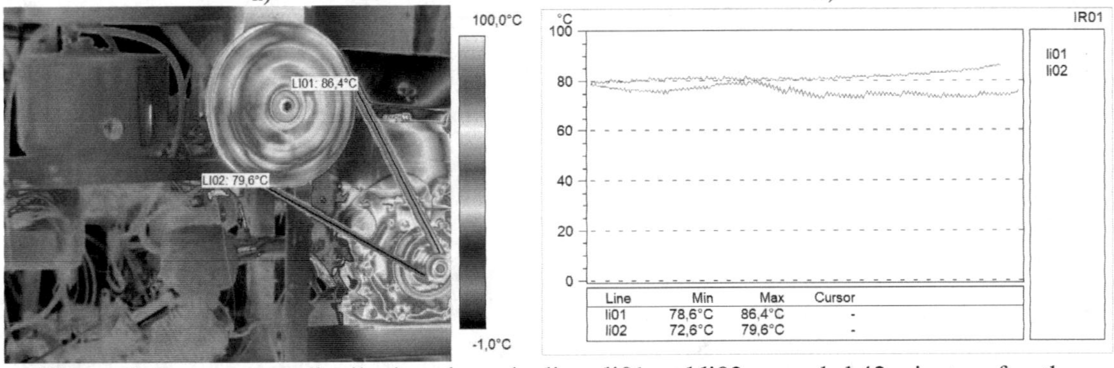

Figure 9. Temperature distribution along the lines li01 and li02 recorded 42 minutes after the commencement of comminution, (two operators, the distance of the branch from the chipper 9 metres) red - active cable, green - passive cable.

In the figure (Fig. 9a) please see the terminogram made on the playback side 42 min. You can then turn on the stabilization of temperature changes. The temperature of the tendon joining the active pulley (82.5 °C), while the temperature of the passive cable fluctuated around the value (76.1 °C). At that time, the temperature difference in the tendons between the active (82.5 °C) and passive tendons was (approx. 6.4 °C). In (Fig. 9b) there are profilograms describing this condition of the transmission, the stabilization of temperature distribution is visible.

The thermographic image (Fig. 10a) observed during the experiment at a distance of 15 meters, branch diameter approx. 100 mm, is very interesting due to the analyzed aspects of the operation, as well as for engineering practice. The temperature distribution at several belt measurement points was measured and the temperature of the passive cable was found to be lower than that of the active cable. Between the temperature of the active (71.25 °C) and passive (63.8 °C) cables, the difference was ΔT = 7.45 °C (Fig. 10). Such a behavior of the transmission under the export conditions tested is caused by cooling down due to transverse vibrations of the belt.

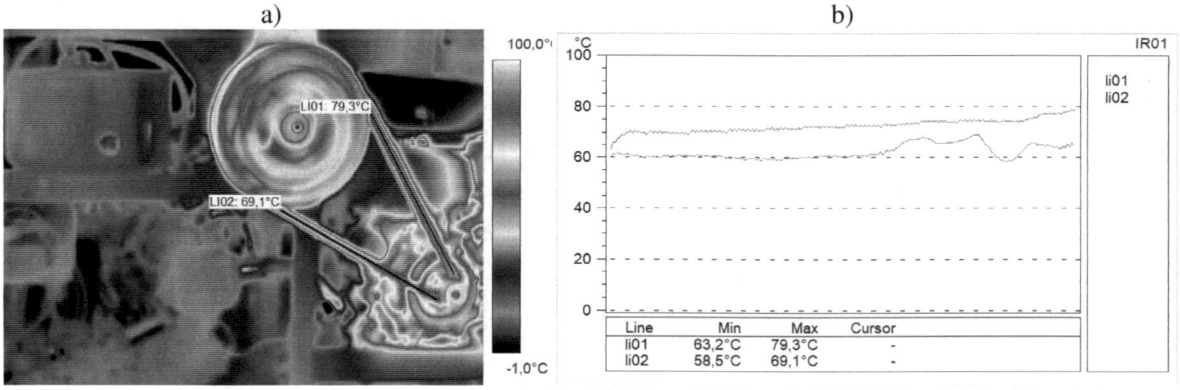

Figure 10. Temperature distribution along the lines li01 and li02 recorded 3.35 minutes after the commencement of comminution, (one operator, the distance of the branch from the chipper 15 metres) red - active cable, green - passive cable.

Figure 11 shows changes in V-belt temperature and profilograms recorded during the third stage of the tests after 24 minutes at a distance of 15 meters, branches with a diameter of about 100 mm were fed. After the first 30 kg of chips were produced, a very rapid increase in the average temperature was observed in the active cable (78.08 °C) in the entry phase and in the passive section of the belt (72.15 °C) during its exit from the driving pulley.

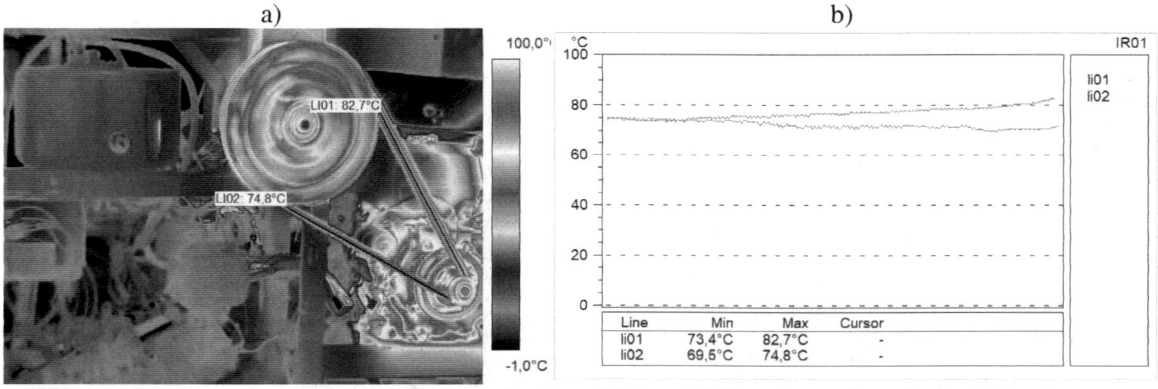

Figure 11. Temperature distribution along the lines li01 and li02 registered, registered 24 minutes from the commencement of comminution (two operators, the distance from the branch to the chipper 15 metres), red - active cable, green - passive cable.

IOP Conf. Series: Materials Science and Engineering 1016 (2021) 012013 doi:10.1088/1757-899X/1016/1/012013

In the tested case, under the given conditions of the experiment, an increase in the temperature difference between the temperature values in the active and passive cable was observed, corresponding to the value of ΔT= 5,93 °C. Such a high temperature of the belts was caused by the operation of the device during very unfavorable working conditions.

Figure 12 shows the temperature distribution of the belt and the profilograms recorded during the third stage of the tests, after 48 minutes at a distance of 15 meters, the branches with a diameter of about 100 mm were fed. After producing 261 kg of chips, a very rapid increase in the mean temperature was observed in the active cable (75.945 °C) in the relaxation phase and in the passive cable (62.5 °C) in the drop-off phase from the driving wheel. In the tested case, under the given conditions of the experiment, an increase in the temperature difference between the temperature values in the active and passive passive was observed, corresponding to the value of ΔT= 12.95 °C. Such a high temperature of the belts was caused by the operation of the device during very unfavorable working conditions.

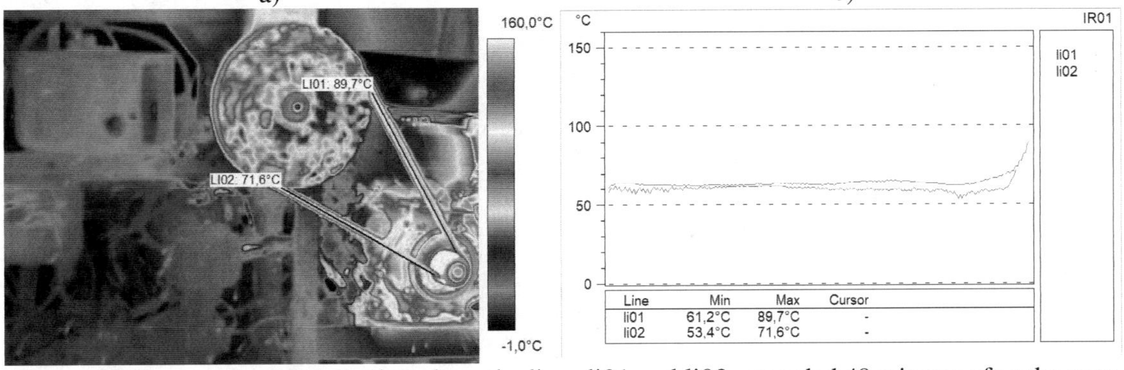

Figure 12. Temperature distribution along the lines li01 and li02 recorded 48 minutes after the commencement of comminution (two operators, the distance of the branches from the chipper 15 metres (only by resistance to gear movement), red - active cable, green - passive cable.

The last thermogram was recorded after the belt was broken. It took place immediately after recording the thermogram presented in Figure 13. It was caused by a very fast increase in belt temperature caused by blockage of the shredder.

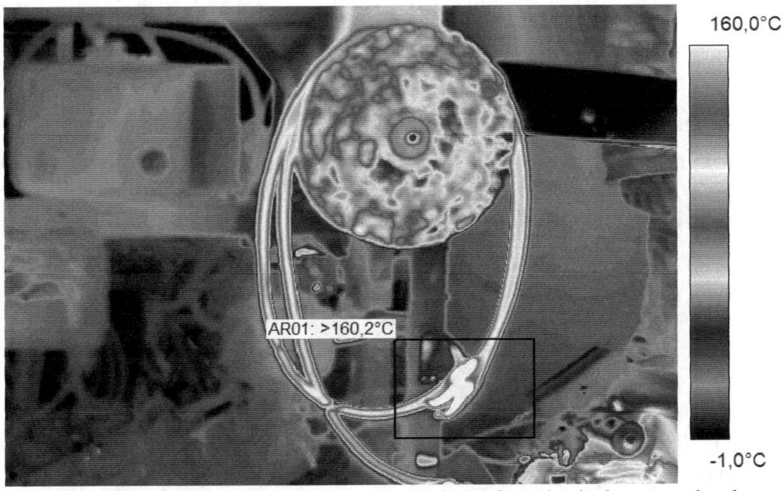

Figure 13. The thermogram recorded after the belts were broken

Figure 14 shows the relation of belt temperature change to capacity for a branch distance of 3 m from the chipper.

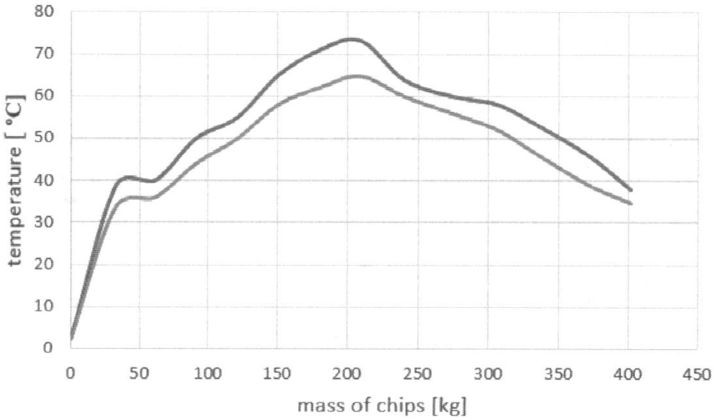

Figure 14. Relation of belt temperature change to efficiency for a branch distance of 3 m from the chipper, red - active cable, green - passive cable

Conclusions

The paper presents the results of thermal imaging tests of a tension transmission with a V-belt used to drive a wood chipper. The correctness of gear operation was assessed on the basis of temperature changes on the belt surface. According to the authors, there is a relationship between the characteristics of the friction process, transmitted power, efficiency, load and material properties of the V-belt, etc. Summarizing the observation results, it should be stated that the thermal imaging camera is an appropriate tool to assess the warm state of the V-belt transmission used in wood chippers. Based on the tests, it can be concluded that: for a distance of 3 meters from the shredder, at the initial stage and in the final operation of the machine, a stabilized temperature was recorded in the active cable of approx. 38 °C and the passive of 34 °C. A large increase in temperature was noted when working with two operators, then the temperature in the active cable was about 73 °C and the passive cable temperature was 65 °C. This was due to the high load on the transmission and its continuous operation without idling. For a distance of 9 meters from the shredder, a similar temperature of about 80 °C was recorded in the active cable, and in the passive of 74 °C in the entire time interval. The reason for maintaining such a high temperature was the large cross-section of branches of approx. 100 mm (constituting the limit load of the machine).

For a distance of 15 meters from the shredder, a gradual increase in temperature in the active cable was recorded. Despite the greater distance from the machine, the temperature rises dangerously, approaching the limit values given in the catalogues of belt manufacturers. It is essential to choose a belt for such difficult working conditions. Figure 13 shows the final stage of the operation of the gear subjected to a very high load. This caused the temperature of the belt to rise to about 160 °C and consequently its breaking. High temperatures may cause the belts to merge or break [36, 37].

On the basis of the conducted tests and analyzes, it can be stated that in order to ensure failure-free operation of the wood chipper, one should strive to unify the temperature in the entire V-belt. The conducted experimental studies to assess the thermal condition of mechanical devices confirmed the usefulness of thermography in the diagnosis of the technical condition of cable gears with a classic V-belt, and the obtained results will certainly be useful for the designers using gears [38-40] and the people dealing with maintenance in enterprises.

References

[1] Krawiec P 2010 Numerical Analysis of Geometrical Characteristics of Machine Elements Obtained Through CMM Scanning *Progress in Industrial Mathematics* Springer-Verlag, Berlin–Heidelberg 925-30

[2] Kujawski M and Krawiec P 2011 Analysis of Generation Capabilities of Noncircular Cog belt Pulleys on the Example of a Gear with an Elliptical Pitch Line *Journal of Manufacturing Science and Engineering-Transactions of the ASME* **133**(5) 051006

[3] Krawiec P Grzelka M Kroczak J Domek G and Kołodziej A 2019 A proposal of measurement methodology and assessment of manufacturing methods of nontypical cog belt pulleys *Measurement* **132** 182-90
[4] Krawiec P and Marlewski A 2011 Spline description of not typical gears for belt transmissions. *Journal of Theoretical and Applied Mechanics* **49**(2) 355-67
[5] Krawiec A and Marlewski A 2016 *Profile design of noncircular belt pulleys Journal of Theoretical and Applied Mechanics* **54** (2) 561-70
[6] Krawiec P 2017 Analysis of selected dynamic features of a two-whwwled transmission system *Journal of Theoretical and Applied Mechanics* **55** (2) 461-7
[7] Krawiec P Waluś K Warguła Ł and Adamiec J 2018 Wear evaluation of elements of V-belt transmission with the application of optical microscope. *MATEC Web of Conferences* **157** 01009-1-8 https://doi.org/10.1051/matecconf/201815701009
[8] Krawiec P Waluś K J Warguła Ł and Adamiec J 2019 Wear evaluation study of the multiple grooved pulleys with optical method *MATEC Web of Conferences* **254** 01004-1-8 https://doi.org/10.1051/matecconf/201925401004
[9] Krawiec P Różański L Czarnecka-Komorowska D and Warguła Ł 2020 Evaluation of the Thermal Stability and Surface Characteristics of Thermoplastic Polyurethane V-Belt *Materials* **3**(7) 1502 https://doi.org/10.3390/ma13071502-25
[10] Krawiec P Warguła Ł Dziechciarz A Małozięć D and Ondrušová D 2020 Evaluation of chemical compound emissions during thermal decomposition and combustion of V-belts *Przemysł Chemiczny* **99** (1) 92-8
[11] Czarnecka-Komorowska D and Wiszumirska K 2020 Sustainability design of plastic packaging for the Circular Economy *Polimery* **65** 8-17
[12] Czarnecka-Komorowska D Wiszumirska K and Garbacz T 2018 Films LDPE/LLDPE made from post-consumer plastic processing structure mechanical properties *Adv. Sci. Technol. Res. J.*, **12** 134-1
[13] Czarnecka-Komorowska D Sterzynski T Maciejewski H and Dutkiewicz M 2012 The effect of polyhedral oligomeric silsesquioxane (POSS) on morphology and mechanical properties of polyoxymethylene (POM) *Composites Theory and Practice* **12** (4) 232-6
[14] Czarnecka-Komorowska D and Mencel K 2014 Modification of polyamide 6 and polyoxymethylene with [3-(2-aminoethyl) amino] propyl-heptaisobutyl-polysilsesquioxane nanoparticles *Przemysł Chemiczny* **93** (3) 392-6
[15] Domek G and Wilczyński M 2019 Modelling a timing belt pitch *MATEC Web of Conferences* **254** 01011
[16] Domek G Kołodziej A Dudziak M and Woźniak T 2017 Identification of the quality of timing belt pulleys *Procedia Engineering* **177** 275-80
[17] Domek G Kołodziej A Wilczyński M and Krawiec P 2017 The problem of cooperation of a flat belts with elements of mechatronic systems *55th International Scientific Conference on Experimental Stress Analysis* 706-711
[18] Łazarska M Woźniak T Ranachowski Z Trafarski A and Domek G 2017 Analysis of acoustic emission signals at austempering of steels using neural networks *Metals and Materials International* 23 3 pp 426-433
[19] Wilczyński M and Domek G 2019 Influence of tension layer quality on mechanical properties of timing belts *MATEC Web of Conferences* **254** 05010
[20] Dudek K Banasiak J and Bieniek J 2003 Diagnosis of kinematic pairs in the rotary mowers Eksploatacja i Niezawodność – *Maintenance and Reliability Eksploatacja i niezawodność* **20**(4) 17-21
[21] Silva C A F Manin L Rinaldi R G Remond D Besnier E and Andrianoely M A 2018 Modeling of power losses in poly-V belt transmissions: hysteresis phenomena (enhanced analysis*) Mech. Mach. Theory* **121** 373-97
[22] Stawicki T and Sędłak P 2015 Thermal Imaging Studies on Grain Harvester Belt Transmissions *Diagnostyka* **16** (2) 37-42

IOP Conf. Series: Materials Science and Engineering 1016 (2021) 012013 doi:10.1088/1757-899X/1016/1/012013

[23] Sundararaman S Hu J Chen J and Chandrashekhara K 2009 Temperature dependent fatigue-failure analysis of V-ribbed serpentine belts *Int. J. Fatigue* **31**(8) 1262-70
[24] Jia S S and Song Y M 2015 Elastic dynamic analysis of synchronous belt drive system using absolute nodal coordinate formulation *Nonlinear. Dynam.* **81**(3) 1393-410
[25] Li W B and Xin Z X 2018 Flexural fatigue life prediction of a tooth V-belt made of fiber reinforced rubber *Int. J. Fatigue* **111** 269-77
[26] Choraś R 2005 *Komputerowa wizja Metody interpretacji i identyfikacji obiektów* Warszawa EXIT
[27] Chrzanowski K 2000 Non-contact Thermometry – *Measurement Errors*. Warszawa, Polski Oddział SPIE
[28] Fidali M and Urbanek G 2007 Application of evolutionary algorithm to limitation of a set of statistical features of thermovision images *Computer Assisted Mechanics and Engineering Sciences* **14** (4) 601-9
[29] Gade R Moeslund T 2014 Thermal cameras and applications: a survey. Machine Vision and Applications **25** 245-62
[30] Hooke C J Kukureka SvN Liao P Rao M and Chen Y K 1996 The friction and wear of polymers in non-conformal contacts *Wear* **200** (1-2) 83-94
[31] Nachman M and Kwiatkowski K 2013 The effect of thermal annealing on the abrasion resistance of a segmented block copolymer urethane elastomers *Wear* **306** (1-2) 113-8
[32] Warguła Ł Krawiec P Waluś K J and Kukla M 2020 Fuel Consumption Test Results for a Self-Adaptive Maintenance-Free Wood Chipper Drive Control System *Applied Sciences* **10** 2727
[33] Warguła Ł Kukla M Krawiec P and Wieczorek B 2020 Reduction in Operating Costs and Environmental Impact Consisting in the Modernization of the Low-Power Cylindrical Wood Chipper Power Unit by Using Alternative Fuel *Energies* **13** 2995
[34] Warguła Ł Kukla M Lijewski P Dobrzyński M and Markiewicz F 2020 Influence of Innovative Woodchipper Speed Control Systems on Exhaust Gas Emissions and Fuel Consumption in Urban Areas *Energies* **13** 3330
[35] Warguła Ł Kukla M Krawiec P and Wieczorek B 2020 Impact of Number of Operators and Distance to Branch Piles on Woodchipper Operation *Forests* **11** 598
[36] Wałęsa K Malujda I and Talaśka K 2018 Butt welding of round drive belts *Acta mechanica et automatica* **12**(2) 115-26
[37] Wałęsa K Malujda I Talaśka K and Wilczyński D 2020 Process analysis of the hot plate welding of drive belts *Acta mechanica et automatica* **14**(2) 84-90
[38] Wieczorek B Warguła Ł and Rybarczyk D 2020 Impact of a hybrid assisted wheelchair propulsion system on motion kinematics during climbing up a slope *Applied Sciences* **10**(3) 1025
[39] Wieczorek B and Kukla M 2020 Biomechanical Relationships Between Manual Wheelchair Steering and the Position of the Human Body's Center of Gravity *Journal of Biomechanical Engineering* **142**(8)
[40] Wieczorek B and Kukla M 2019 Effects of the performance parameters of a wheelchair on the changes in the position of the centre of gravity of the human body in dynamic condition *PloS one* **14**(12) e0226013

IOP Conf. Series: Materials Science and Engineering 1016 (2021) 012014 doi:10.1088/1757-899X/1016/1/012014

Development of IoT Weather Monitoring System Based on Arduino and ESP8266 Wi-Fi Module

Andriy Holovatyy

Department of Computer-Aided Design Systems, Lviv Polytechnic National University, 5 Mytropolyta Andreya St., building 4, 79013 Lviv, Ukraine

Email: andrii.i.holovatyi@lpnu.ua

Abstract. In the paper, the hardware and software of the IoT weather monitoring system have been developed. The hardware of the IoT weather monitoring system consists of the microcontroller device which is developed on the Arduino Mega2560 board, digital pressure, temperature and humidity sensor BME280, Wi-Fi module ESP-01 built on the ESP8266 chip and the 16×2 alphanumeric LCD based on the Hitachi HD44780 controller. The electronic circuit and the model of the microcontroller weather monitoring device have been created in Proteus VSM. The operation algorithm of the IoT weather monitoring system has been developed. The IoT device monitors such weather parameters as: atmospheric pressure, temperature and relative humidity. The software modules for communication with the BME280 sensor and Wi-Fi module ESP-01 and main data collection and acquisition software for the ATmega2560 microcontroller of the Arduino Mega2560 board have been created. The software for communication with the ThingsBoard IoT platform and Mosquitto MQTT broker has been created. The dashboards for weather data visualization on the IoT platform ThingsBoard using ThingsBoard Dashboard and from the Mosquitto MQTT broker using Node-RED have been created. The operation of the microcontroller weather monitoring device has been simulated in Proteus ISIS. The prototype of the microcontroller weather monitoring device has been built on a solderless breadboard and the operation of the IoT weather monitoring system has been tested.

1. Introduction

The development of electronics, programming technologies and Internet has led to the emergence of the Internet of Things. The Internet of Things (IoT) is a set of technologies for collecting information from a system of distributed sensors and remotely controlling devices connected to the Internet, as well as for storing, processing and visualizing data on local or remote servers. Devices that are part of the Internet of Things are any stand-alone devices that are connected to the Internet and can be tracked and/or controlled remotely [1-2].

One IoT device connects to another to transmit information over Internet protocols. IoT platforms serve as a bridge between device sensors and the data network [3]. Today, there is a wide variety of IoT platforms, such as: Amazon Web Services, Microsoft Azure, ThingWorx IoT Platform, IBM's Watson, Cisco IoT Cloud Connect, Salesforce IoT Cloud, Oracle Integrated Cloud, GE Predix, Node-RED, ThingsBoard, Adafruit IO , Kaa IoT Platform, thinger.io.

Today, one of the commonly-used data communication protocol for IoT is MQTT. MQTT is a lightweight publish/subscribe messaging protocol designed for M2M (machine-to-machine) telemetry in low-bandwidth environments. MQTT stands for Message Queuing Telemetry Transport, but was previously called Telemetry Transport with a message queue [4].

MQTT (Message Queue Telemetry Transport) protocol is a lightweight messaging protocol for sending simple data flows from sensors to applications, middleware or analytics platforms and cloud solutions. Today, MQTT is the main protocol used for the Internet of Things (IoT). The MQTT publish-subscribe model is shown in Fig. 1.

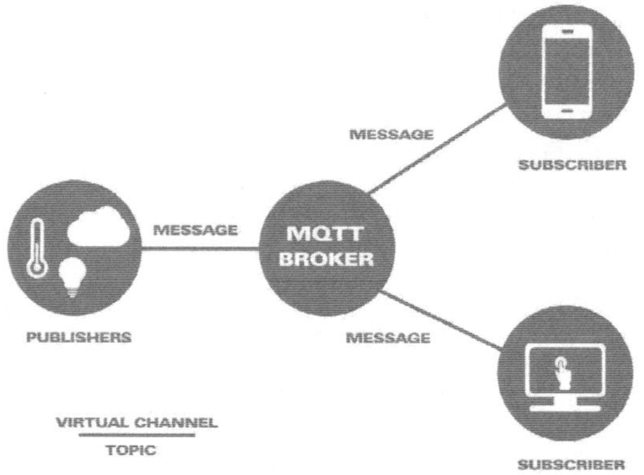

Figure 1. MQTT publish-subscribe model

The purpose of this work is to create hardware and software IoT system for monitoring weather parameters. For the development of the hardware of the IoT weather monitoring system the Arduino Mega2560 board based on the ATmega2560 microcontroller [5], BME280 digital atmospheric pressure, humidity and temperature sensor [6], ESP-01 Wi-Fi module built on the ESP8266 chip [7] and alphanumeric 16×2 LCD module based on the HD44780 controller are used.

The microcontroller device has to measure weather parameters and transmit them over Wi-Fi network using the MQTT protocol to the Mosquitto server (broker) [8] or to the ThingsBoard IoT platform [9]. The ThingsBoard IoT platform collects, processes the received IoT data and displays them on the created weather data visualization dashboard using ThingsBoard Dashboard. The Mosquitto server (broker) sends the weather data to the MQTT client subscribed to the topic. The MQTT client displays them on the monitoring and visualization dashboard created by using Node-RED [10].

Many modern electronic devices are built on microcontrollers [11-16]. The important feature for the designer is that they make possible to implement easier and cheaper the operation algorithm of the device and reduce its dimensions. Using the microcontrollers the data can be easy received from different sensors and different actuators (motors, heaters, lighting devices, servos, etc.) can be controlled, mathematical calculations can be performed.

2. Hardware design of IoT weather monitoring system in Proteus VSM

IoT system has to monitor such weather parameters as: temperature, relative humidity, atmospheric pressure. In Fig. 2, the developed block diagram of the hardware of the IoT weather monitoring system is shown. The hardware of the IoT weather monitoring system is built on the Arduino Mega2560 board based on the ATmega2560 microcontroller. The BME280 digital pressure, temperature and humidity sensor, ESP8266 Wi-Fi module, alphanumeric LCD module HD44780, 4 buttons are connected to the Arduino Mega2560 board. The Arduino ATmega2560 microcontroller reads and processes weather data from the BME280 sensor. The device displays the current values of the weather parameters on the LCD.

IOP Conf. Series: Materials Science and Engineering 1016 (2021) 012014 doi:10.1088/1757-899X/1016/1/012014

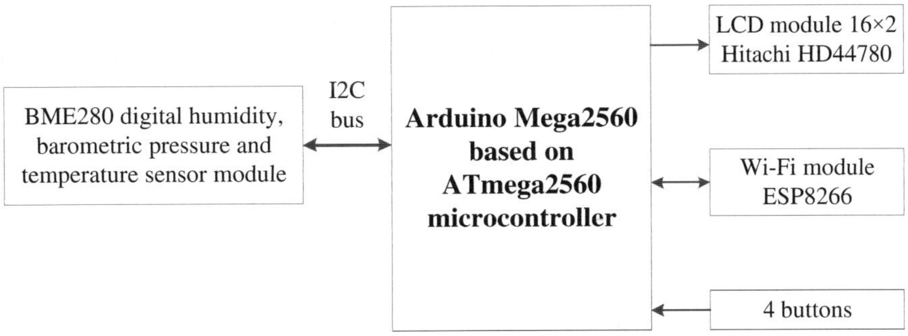

Figure 2. Hardware block diagram of the IoT weather monitoring system (microcontroller device based on Arduino)

In Fig. 3 the developed hardware of the IoT weather monitoring system in the Proteus CAD software is shown. In the circuit, the BME280 digital atmospheric pressure, temperature and humidity sensor is connected to the I2C bus to the pins 20 (SDA/INT1) and 21 (SCL/INT0)) on the Arduino Mega2560 board. The pins ESP_RX and EXP_TX of the ESP8266 Wi-Fi module are connected via the UART interface to the pins 16 (TX2) and 17 (RX2) on the Arduino Mega2560 board. The 16 character×2 line LCD module based on HD44780 controller is connected to the pins 2…7 (PE0/PCINT8/RXD0, PE1/TXD0, PE4/INT4/OC3B, PE5/INT5/OC3C, PG5/OC0B, PE3/AINT PH3/OC4A, PH4/OC4B) on the Arduino Mega2560. The register select pin (RS) of the LCD module is connected to the Arduino Mega2560 digital pin 2. The enable/control pin (E) is connected to the digital pin 3 on the Arduino Mega2560 board. The data pins D4…D7 of the LCD module are connected to the digital pins 4…7 on the Arduino Mega2560 board, respectively. The 10 kOhm RV1 potentiometer is connected to the VEE LCD module pin. This potentiometer is used to supply 0…5V voltage to adjust the display contrast.

The MENU_BTN button is connected to the pin 19 (PD2/INT2/RXD1), the SELECT_PLUS_BTN (select/increase) button to the pin 23 (PA1/AD1), the SELECT_MINUS_BTN button (select/decrease) to the pin 22 (PA0/AD0) and the button to exit from the menu into the operation mode EXIT_BTN to the pin 24 (PA2 / AD2) on the Arduino Mega2560 board.

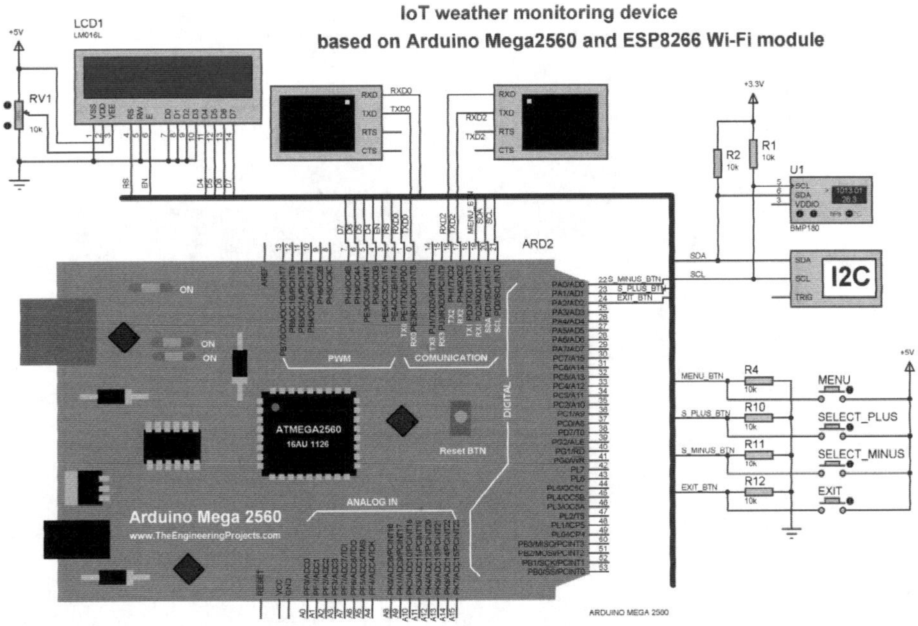

Figure 3. Hardware of the IoT weather monitoring system developed in Proteus VSM

IOP Conf. Series: Materials Science and Engineering 1016 (2021) 012014 doi:10.1088/1757-899X/1016/1/012014

3. Operation algorithm of the IoT weather monitoring system

In Fig. 4 the block diagram of the operation algorithm of the IoT weather monitoring system is shown.

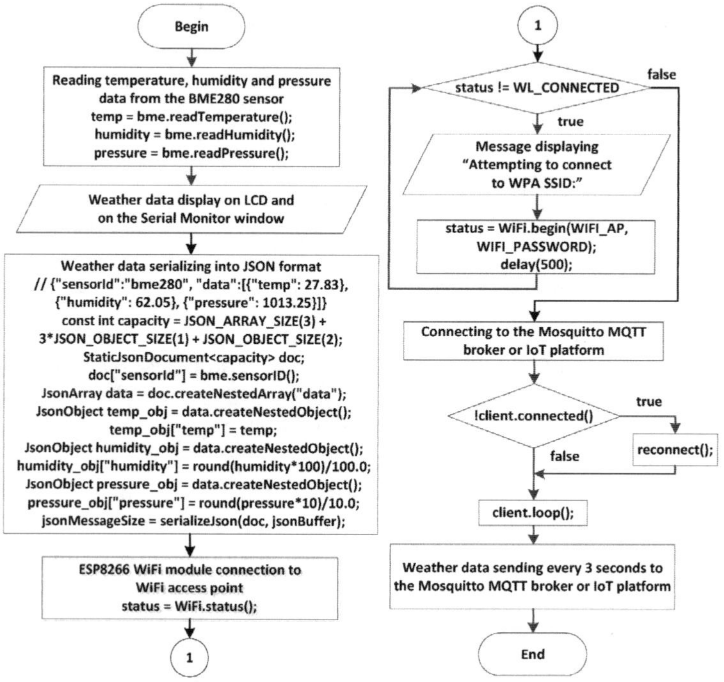

Figure 4. Operation algorithm of the IoT weather monitoring system

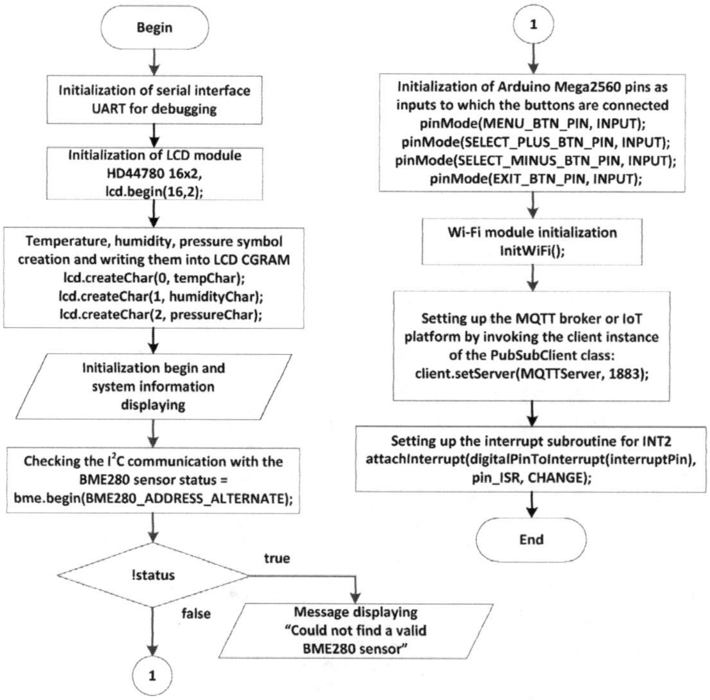

Figure 5. Initialization algorithm of the LCD module, BME280 sensor, ESP8266 Wi-Fi module and MQTT client

After supplying the power to the IoT weather monitoring device, the program written in the microcontroller flash memory starts. First of all, the setup() function initializes the LCD interface and determines the display size of 16 columns and 2 rows, then initializes the serial interface and sets up its speed in bits per second (bauds) (Fig. 5).

The next step is the initialization of the BME280 atmospheric pressure, temperature and humidity sensor connected via the I2C interface with the address 0×44. The initialization of the pins as inputs to which the MENU_BTN, SELECT_PLUS_BTN, SELECT_MINUS_BTN and EXIT_BTN buttons are connected:

Next, the ESP8266 Wi-Fi module is initialized as the MQTT client, the setting up the MQTT server and the connection port for messaging by using the setServer method of the PubSubClient object.

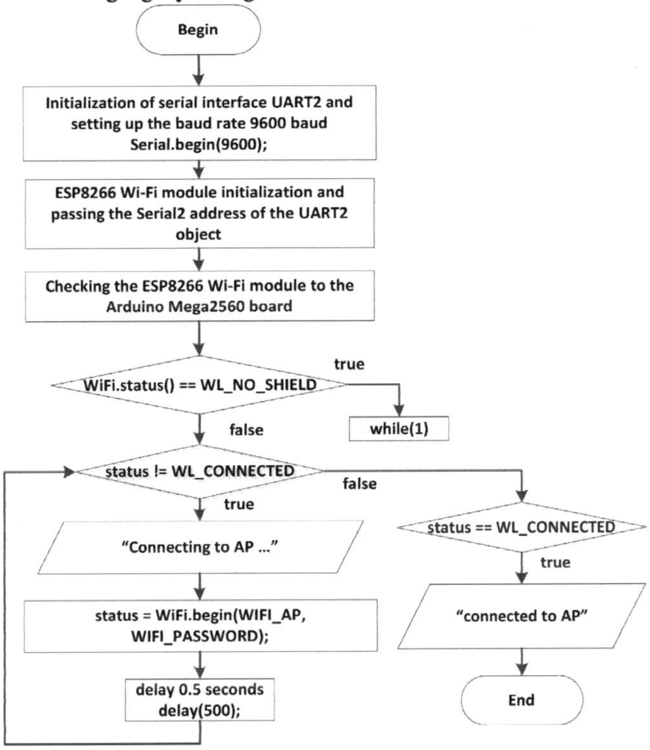

Figure 6. Algorithm of initializing the ESP8266 Wi-Fi module and setting up the connection to AP

Then, the measuring loop of the weather parameters starts, their values transmit to the MQTT server and the measured weather data are displayed on the LCD module.

```
// reading temperature, humidity and pressure from the BME280 sensor
  temp = bme.readTemperature();
  lcd.setCursor(0, 0); lcd.write((byte) 0); lcd.print(' ');
  lcd.print(temp,1); lcd.print(char(223)); lcd.print("C ");
  Serial.print("Temperature = ");
  Serial.print(temp); Serial.write(176); Serial.println("C");
  humidity = bme.readHumidity();  lcd.write((byte) 1); lcd.print(' ');
  lcd.print(humidity, 1);  lcd.print(F("% RH "));
  Serial.print("Humidity = "); Serial.print(humidity);Serial.println(" %");
  pressure = bme.readPressure() / 100.0F * 0.7500616827;
  lcd.setCursor(0, 1); lcd.write((byte) 2); lcd.print(' ');
  lcd.print(pressure, 1); lcd.print(" mmHg");
  Serial.print("Atmospheric pressure = ");
  Serial.print(pressure); Serial.println(" mmHg");
  altetude = bme.readAltitude(SEA_LEVEL_PRESSURE_HPA);
  Serial.print("Altitude = ");
  Serial.print(altetude); Serial.println(" m"); Serial.println();
```

After collecting data, the device connects to the Wi-Fi access point and sends weather data to the MQTT server as MQTT message in JSON format. MQTT messages are published to the appropriate topic using the publish() method, which takes as parameters a topic and message. Next, attempting the connection to the MQTT server in the loop until the connection to the server is successful:

```
char jsonBuffer[512];
size_t jsonMessageSize;

if (!mqtt_server_ind) {
   // {"sensorId":"bme280", "data":[{"temp": 27.83}, {"humidity": 62.05},
{"pressure": 1013.25}]}
   const int capacity = JSON_ARRAY_SIZE(3) + 3*JSON_OBJECT_SIZE(1) +
JSON_OBJECT_SIZE(2);
   DynamicJsonDocument doc(capacity);
   doc["sensorId"] = bme.sensorID();
   JsonArray data = doc.createNestedArray("data");
   JsonObject temp_obj = data.createNestedObject();
   temp_obj["temp"] = temp;
   JsonObject humidity_obj = data.createNestedObject();
   humidity_obj["humidity"] = round(humidity*100)/100.0;
   JsonObject pressure_obj = data.createNestedObject();
   pressure_obj["pressure"] = round(pressure*10)/10.0;
   serializeJson(doc, jsonBuffer);
  }
  else {
    // {"temp": 27.83, "humidity": 62.05, "pressure": 1013.25}
    const int capacity = JSON_OBJECT_SIZE(3);
    DynamicJsonDocument doc(capacity);
    doc["temp"] = temp;
    doc["humidity"] = round(humidity*100)/100.0;
    doc["pressure"] = round(pressure*10)/10.0;
    serializeJson(doc, jsonBuffer);
  }
  Serial.println("Sending message to MQTT topic..");
  Serial.println(jsonBuffer);
…..
  if (!client.connected() ) {
    reconnect();
  }
  if ( millis() - lastSend >= 1000 ) { // data update and sending
     if (client.publish(topic[mqtt_server_ind], jsonBuffer,
jsonMessageSize) == true) {
        Serial.println("Success sending message");
     } else { Serial.println("Error sending message"); }
     lastSend = millis();
 }
  client.loop();
```

The interrupt subroutine ISR_BtnPressed() is called when the external interrupt INT2 occurs. The external interrupt is generated when the MENU_BTN button, connected to the pin 19 (INT2), is pressed. The interrupt is generated each time the pin output state changes - CHANGE mode.

```
void ISR_BtnPressed() {
  if (digitalRead(MENU_BTN_PIN) == HIGH)   {
     detachInterrupt(digitalPinToInterrupt(interruptPin));
     mainMenuInvoke = true;
  }
}
```

In the main loop, only pressing the MENU_BTN button is processed. After pressing it, the main menu is invoked. In the main menu a user can set up the IoT weather monitoring device:

IOP Conf. Series: Materials Science and Engineering 1016 (2021) 012014 doi:10.1088/1757-899X/1016/1/012014

4. Simulation and prototype testing of the IoT weather monitoring system

The written sketch (program) in Arduino IDE is compiled and linked to the hex firmware file for the ATmega2560 microcontroller of the Arduino Mega2560 board. Before downloading the firmware to the Arduino board, the correctness of its operation on the created device model in Proteus is checked. The Proteus VSM software makes possible to simulate and debug quite complex devices. Such devices can have several microcontrollers simultaneously and even from different families. In the figures below the simulation of the IoT weather monitoring system in Proteus ISIS is shown. In Fig. 7 the simulation of IoT weather monitoring device in Proteus ISIS is shown.

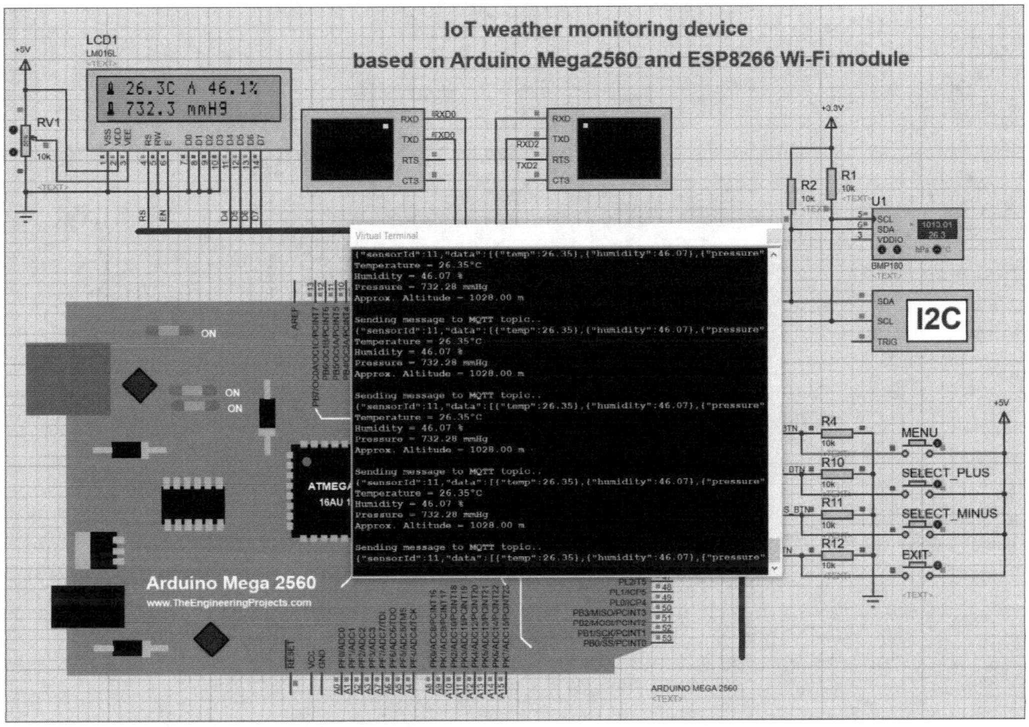

Figure 7. Simulation of the IoT weather monitoring device in Proteus ISIS

In Fig. 8-17 the prototype testing of the IoT weather monitoring system is shown.

Figure 8. Test circuit (prototype) of the IoT weather monitoring device built on the Arduino Mega2560 board, BME280 digital atmospheric pressure, temperature and humidity sensor, ESP8266 Wi-Fi module

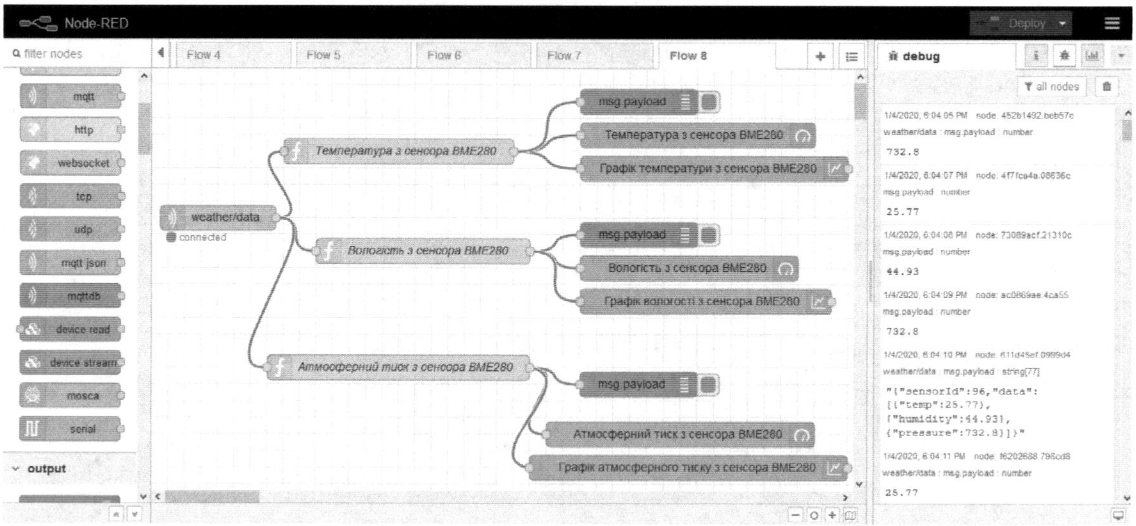

Figure 9. Node-RED dashboard flow for the IoT weather data visualization

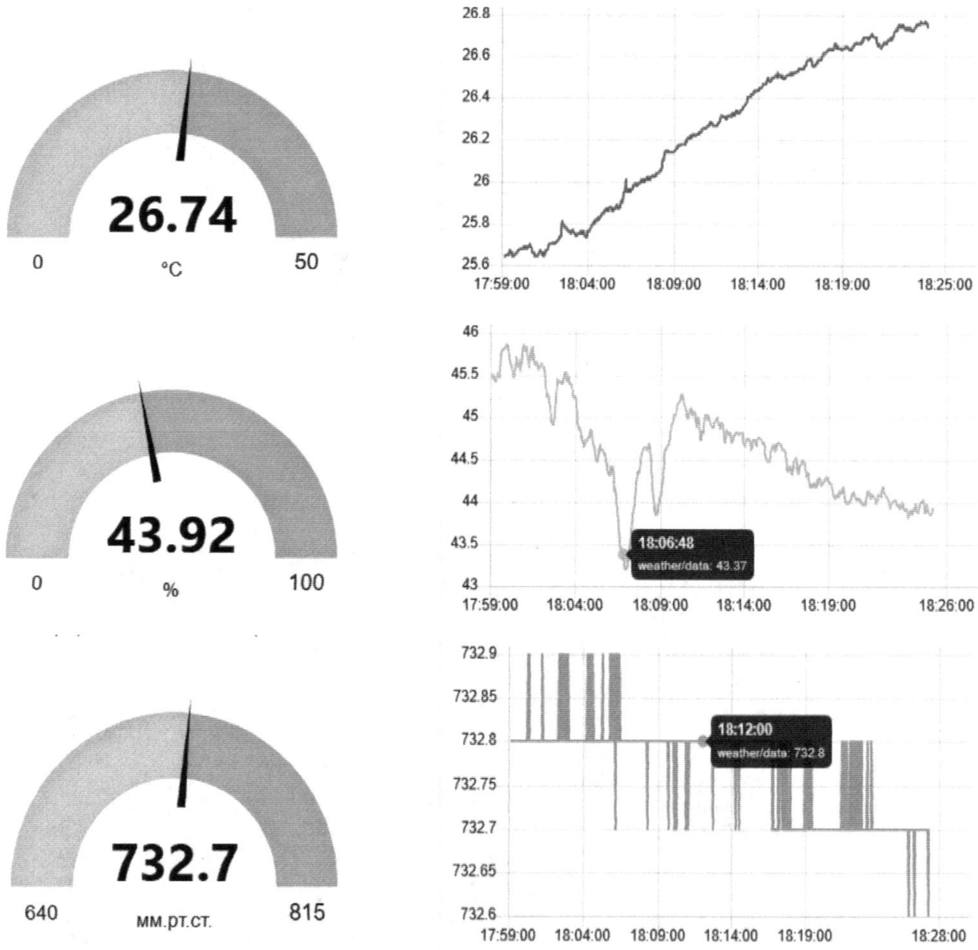

Figure 10. IoT weather data visualization on the Node-RED Dashboard

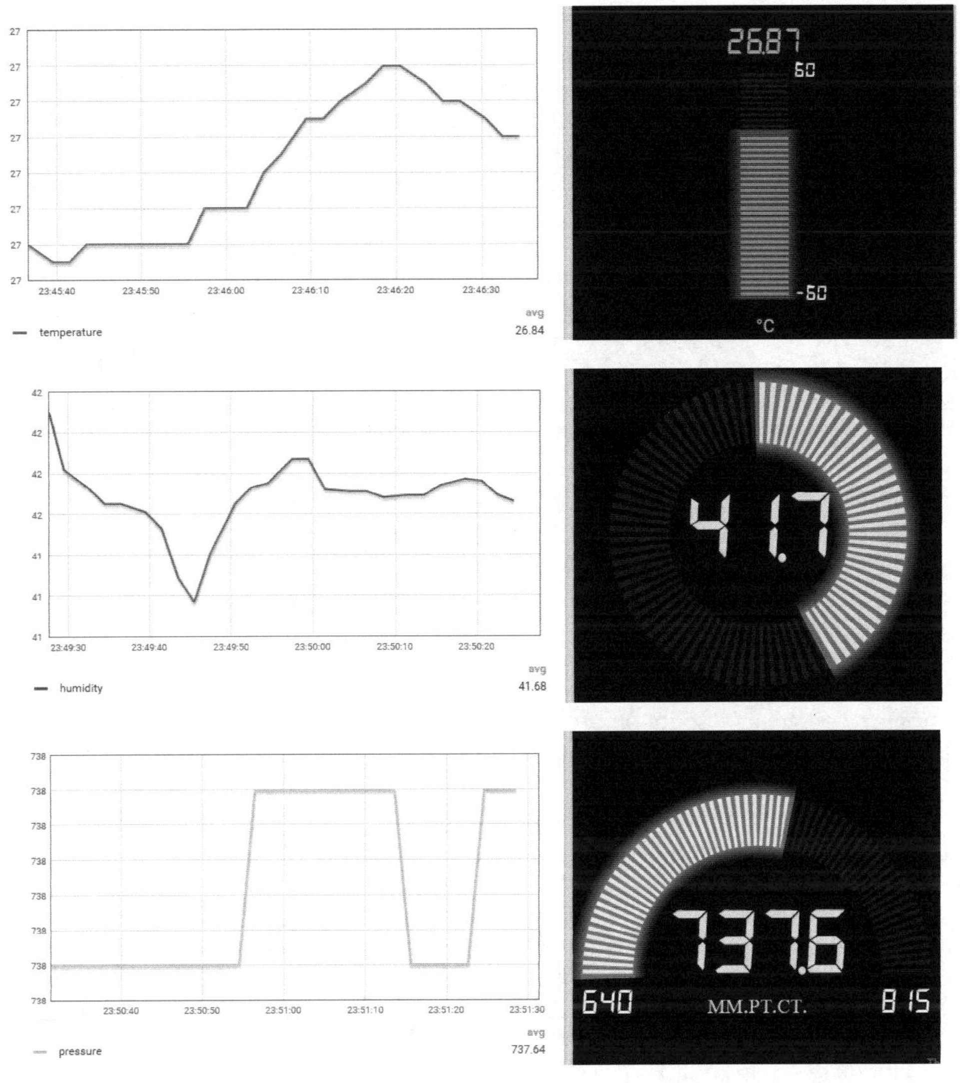

Figure 11. IoT weather data visualization on ThingsBoard Dashboard

Conclusions

In the work, the hardware and software of the IoT weather monitoring systems have been developed. The hardware of the IoT weather monitoring system consists of the microcontroller device that measures weather parameters such as temperature, relative humidity, atmospheric pressure and sends them to the IoT platform for further processing and visualization.

The microcontroller device is built on the Arduino Mega2560 board based on the ATmega2560 microcontroller, the BME280 digital atmospheric pressure, temperature and humidity sensor, ESP-01 Wi-Fi module based on the ESP8266 chip and 16×2 HD44780 LCD module. The microcontroller-based device for weather monitoring has the menu that allows a user to select Wi-Fi access point, MQTT server (IoT platform) and data format.

The electronic circuit and model of the IoT weather monitoring device in Proteus VSM have been developed. The operation algorithm and software of the IoT weather monitoring device have been developed. The software consists of modules for communicating with the BME280 digital sensor, ESP-01 Wi-Fi module and the main program for the ATmega2560 microcontroller of the Arduino Mega2560 board. The main program collects, processes weather data and sends them via MQTT protocol to the Mosquitto server or to the ThingsBoard IoT platform. The dashboards for visualization of the received IoT data from the Mosquitto server using Node-RED, and on the ThingsBoard IoT platform using ThingsBoard Dashboard are created.

The operation of the IoT device in the Proteus ISIS emulator has been simulated. The test circuit of the IoT weather monitoring device on the solderless breadboard has been built and its operation has been tested. The results of simulation and testing of the IoT weather monitoring system have showed the correct operation of the developed hardware and software.

References

[1] Peter Waher Pradeeka Seneviratne Brian Russel and Drew Van Duren 2016 *IoT: Building Arduino-Based Projects* (Packt Publishing) p732
[2] Adeel Javed 2016 *Building Arduino Projects For The Internet Of Things: Experiments With Real-World Applications* p 285
[3] Rajesh Singh Anita Gehlor Lovi Raj Gupta Bhupendra Singh and Mahendra Swain 2020 *Internet of Things with Raspberry Pi and Arduino* (CRC Press Taylor & Francis Group LLC) p 190
[4] Electronic resources on MQTT protocols: http://mqtt.org/documentation, http://www.steves-internet-guide.com/mqtt-protocol-messages-overview/
[5] Tim Pulver 2019 *Hands-On Internet of Things with MQTT:Build connected IoT devices with Arduino and MQ Telemetry* (Transport (MQTT) – Packt Publishing 1 edition October 4) p 350
[6] Electronic resource on datasheet of the humidity, pressure and temperature sensor BME280: http://www.embeddedadventures.com/datasheets/BME280.pdf
[7] Electronic resource on description of the ESP8266 Wi-Fi module connection to the Arduino board and its programming: http://wiki.amperka.ru/%D0%BF%D1%80%D0%BE%D0%B4%D1%83%D0%BA%D1%82%D1%8B:esp8266-wifi-module
[8] Electronic resource on the ThingsBoard IoT platform: https://thingsboard.io/docs/user-guide/install/installation-options/?ceInstallType=liveDemo
[9] Electronic resource on MQTT broker Mosquitto: https://mosquitto.org/
[10] Electronic resource on Node-RED: https://nodered.org/docs/getting-started/windows
[11] Iwaniec M Holovatyy A Teslyuk V Lobur M Kolesnyk K and Mashevska M 2017 Development of vibration spectrum analyzer using the Raspberry Pi microcomputer and 3-axis digital MEMS accelerometer ADXL345 *XIIIth International Conference on Perspective Technologies and Methods in MEMS Design (MEMSTECH)* 25-9 doi: 10.1109/MEMSTECH.2017.7937525
[12] Holovatyy A Teslyuk V Lobur M Pobereyko S and Sokolovsky Y 2018 Development of Arduino-Based Embedded System for Detection of Toxic Gases in Air *IEEE 13th International Scientific and Technical Conference on Computer Sciences and Information Technologies (CSIT)* 139-42 doi: 10.1109/STC-CSIT.2018.8526672
[13] Holovatyy A Teslyuk V Lobur M Sokolovskyy Y and Pobereyko S 2018 Development of Background Radiation Monitoring System Based on Arduino Platform *IEEE 13th International Scientific and Technical Conference on Computer Sciences and Information Technologies (CSIT)* 121-4 doi: 10.1109/STC-CSIT.2018.8526696
[14] Kosobutskyy P and Ferens R 2017 Statistical analysis of noise measurement system based on accelerometer-gyroscope GY-521 and Arduino platform *14th International Conference The Experience of Designing and Application of CAD Systems in Microelectronics (CADSM)* 405-7 doi: 10.1109/CADSM.2017.7916162
[15] Lobur M Korpyljov D Jaworski N Iwaniec M and Marikutsa U 2020 Arduino Based Ambient Air Pollution Sensing System *IEEE XVIth International Conference on the Perspective Technologies and Methods in MEMS Design (MEMSTECH)* 32-5 doi: 10.1109/MEMSTECH49584.2020.9109460
[16] Matviykiv O et al 2019 Lab-chip Diagnostic Device for the Rainwater Monitoring System Using Wireless Sensors Network *MIXDES - 26th International Conference "Mixed Design of Integrated Circuits and Systems* 241-5 doi: 10.23919/MIXDES.2019.8787185

IOP Conf. Series: Materials Science and Engineering 1016 (2021) 012015 doi:10.1088/1757-899X/1016/1/012015

Simulation of the Processes in Electrical Engineering Systems via the Two-Point Problem for Telegraph Equation

Z Nytrebych[1] and O Malanchuk[2]*

[1]Lviv Polytechnic National University, 12, Bandery str., Lviv, 79013, Ukraine
[2]Danylo Halytsky Lviv National Medical University, 69, Pekarska str., Lviv, 79017, Ukraine

*Email: oksana.malan@gmail.com

Abstract. The physical-mathematical model which describes distribution of an electric current in electrotechnical systems is investigated. This model contains a homogeneous partial differential equation of the second order in time and two-point time conditions. The coefficients of equation are the parameters of the system. Two-point conditions describe the behavior of the process at two points in time. The analytical solution of the two-point problem which is obtained by the differential-symbol method determines the behavior of the process at any time and at arbitrary point of the system. The class of functions in which exists a unique solution is specified. The presented results can be used in the design of electrical engineering systems and to predict the results in adjusting parameters.

1. Introduction

One of the main functions of the elements of electrical systems is the transmission and distribution of electricity with minimal losses and a high level of reliability. The development of methods and means of controlling the modes of operation of perspective power systems elements facilitates the progress of technology. This can be achieved via mathematical modeling, which allows engineers to verify the compliance of system elements with the technical requirements of the project, using virtual rather than physical experiments. The virtual design methods significantly shorten the design cycle and reduces the cost of design. Obtaining a negative result of the experiment encourages the search for alternatives to create a more effective final project [1]. Modern electrical system is a complex set of elements. Its operation at any time is characterized by a certain value of voltage at different nodes, current in different elements, frequency and other values [2]. The state of the system is determined by changing these parameters. During emergencies, these parameters change very quickly and reach emergency values. It can damage expensive electrical equipment, which will negatively affect the state of the system [3]. Therefore, mathematical modeling, in particular, of electromagnetic processes in electrical systems is a relevant topic for their study to further establish the mechanism of their changes [4, 5].

Many works [6-8] are devoted to the study of processes in electrical systems by methods of mathematical modeling. Construction of a mathematical model which describes electromagnetic and electromechanical components of transients is carried out in [9]. It allows to obtain the distribution of current and makes possible to design electrical networks of arbitrary configuration. In [10] a model of a power line in single-phase form is developed. Special attention is paid to the calculation of mutual capacitances of conductors which determine the energy leakage between conductors and ground, voltage and potential of electrostatic and magnetic fields generated by currents.

IOP Conf. Series: Materials Science and Engineering 1016 (2021) 012015 doi:10.1088/1757-899X/1016/1/012015

2. Topicality

Let consider a two-wire line [11]. Since the magnetic field occurs in a conductor in which an electric current flow, the capacitance and inductance can be determined for each segment of the line. In addition, conductors have non-zero resistance and the layer between them has non-zero conductivity. The distribution of electric current at any time point t and at arbitrary point of a conductor is given by function $i(t,x)$ which is a solution of the telegraph equation, i.e. the partial differential equation

$$\frac{\partial^2 i}{\partial x^2} = CL\frac{\partial^2 i}{\partial t^2} + (g_0 L + CR)\frac{\partial i}{\partial t} + g_0 Ri\,, \tag{1}$$

where R is conductor resistance, L is inductance, g_0 is conductance, C is capacitance [12]. These parameters determine the performance of a line. In [12], partial cases of the telegraph equation for a finite and infinite conductor is studied for different variants of a line, in particular, for a line without leakage (resistance is almost insignificant and insulation is quality, $R = g_0 = 0$), for a line without distortion ($RC = Lg_0$) and others.

The telegraph equation is also used to model the processes of signal propagation [13], electromagnetic wave propagation [14, 15], inverse scattering [16], etc.

In the presented work the problem for the telegraph equation with conditions in which current distribution in an infinite conductor is known in two time moments, instead of at one moment as in Cauchy problem, is investigated. For this problem a differential-symbol method which is convenient to study such problems is used [17]. Applying this method the conditions of existence and uniqueness of solution of the homogeneous two-point in time problem [18] two-point problem for homogeneous equation with nonhomogeneous conditions [19] and two-point problem for nonhomogeneous equation with homogeneous conditions [20, 21] are obtained.

3. Two-point problem for telegraph equation

Let consider a mathematical model of describing distribution of voltage or current in an infinite conductor, which is given by the function $u = u(t,x)$. This function is the solution of the telegraph equation, that is the solution of partial differential equation

$$\left[\frac{1}{\alpha^2}\frac{\partial^2}{\partial t^2} + D\frac{\partial}{\partial t} + E - \frac{\partial^2}{\partial x^2}\right]u = 0, \quad x \in \mathbb{R}, \ t > 0\,. \tag{2}$$

The coefficients α, D, C of equation (2) specify the process parameters. Equation (1) can be obtained from (2) for $\alpha = \frac{1}{\sqrt{LC}}$, $D = Lg_0 + RC$, $E = Rg_0$.

Telegraph equation (2) makes possible to take into account the existing resistances of the medium and associated with them the nature of the wave attenuation. In addition to the processes in electrical systems, equation (2) also describes an elastic displacements of a rope when winding it with load-lifting mechanisms on the drum [22]. Note that the problem for equation (2) with given initial and boundary conditions is investigated in [23, 24]. The solutions of these problems are represented in the integral form using the Riemann functions.

In presented paper, the set of solutions of equation (2) which satisfy two-point time conditions

$$u(0,x) = \phi_1(x), \quad u(\tau,x) = \phi_2(x), \quad x \in \mathbb{R}, \tag{3}$$

is studied. In conditions (3), the functions $\phi_1(x)$ and $\phi_2(x)$ describe the current distribution at two time points $t = 0$ and $t = \tau$, $\tau > 0$. The Cauchy problem for telegraph equation (2) with conditions

$$u(0,x) = \phi_1(x), \quad \frac{\partial u}{\partial t}(0,x) = \phi_2(x), \quad x \in \mathbb{R}\,,$$

for infinitely long conductor is studied in [12].

4. On the existence of unique solution of the two-point problem and its construction

To study problem (2), (3) we can use the differential-symbol method which is successfully applied, for example, to study oscillation processes which is described by wave equation with two-point time conditions [25].

Let us write the ordinary differential equation

$$\left[\frac{1}{\alpha^2}\frac{d^2}{dt^2}+D\frac{d}{dt}+E-\mu^2\right]G(t,\mu)=0. \tag{4}$$

Equation (4) is obtained from (2) with the change $\frac{\partial}{\partial t}$ by $\frac{d}{dt}$ and $\frac{\partial}{\partial x}$ by μ, $\mu\in\mathbb{C}$.

The system of functions

$$G_1(t,\mu)=e^{-\frac{D\alpha^2}{2}t}\left\{D\alpha\frac{\sinh\left[\frac{\alpha}{2}\eta(\mu)t\right]}{\eta(\mu)}+\cosh\left[\frac{\alpha}{2}\eta(\mu)t\right]\right\},\quad G_2(t,\mu)=e^{-\frac{D\alpha^2}{2}t}\frac{\sinh\left[\frac{\alpha}{2}\eta(\mu)t\right]}{\frac{\alpha}{2}\eta(\mu)}, \tag{5}$$

where $\eta(\mu)=\left(D^2\alpha^2+4\left(\mu^2-E\right)\right)^{1/2}$ is the normal fundamental system of solutions of equation (4).

By

$$\Delta(\mu)=G_2(\tau,\mu), \tag{6}$$

we denote the value of the function $G_2(t,\mu)$ at the moment of time $t=\tau$.

Let L be the set of zeroes of the function $\Delta(\mu)$, that is

$$L=\left\{\mu\in\mathbb{C}:\ \alpha^2\eta^2(\mu)\tau^2=-4\pi^2k^2,\ k\in\mathbb{N}\right\}. \tag{7}$$

We consider quasipolynomial of the form

$$\varphi(x)=P(x)e^{\beta x}, \tag{8}$$

where $P(x)$ is polynomial, coefficient β in exponent is complex number. By $K_{\mathbb{C}\setminus L}$, we denote a class of quasipolynomials which can be represented as a finite sum of one-term quasipolynomials (8) with arbitrary polynomials $P(x)$ and different coefficients β, moreover these coefficients belong to the set $\mathbb{C}\setminus L$, where L is the set (7).

If the right-hand sides of conditions (3), i.e. the functions $\phi_1(x)$ and $\phi_2(x)$ are quasipolynomials of the class $K_{\mathbb{C}\setminus L}$ [26], then problem (2), (3) has unique solution in the class of quasipolynomials of variables t and x which for each fixed $t>0$ belong to $K_{\mathbb{C}\setminus L}$. This solution can be found by the formula [27]

$$u(t,x)=\phi_1\left(\frac{\partial}{\partial\mu}\right)\left\{G_1(t,\mu)e^{\mu x}\right\}\Big|_{\mu=0}+\phi_2\left(\frac{\partial}{\partial\mu}\right)\left\{G_2(t,\mu)e^{\mu x}\right\}\Big|_{\mu=0}, \tag{9}$$

where

$$G_1(t,\mu)=\frac{2e^{-\frac{1}{2}\alpha^2D(t+\tau)}}{\Delta(\mu)}\left\{\frac{\sinh\left[\frac{1}{2}\alpha(\tau-t)\eta(\mu)\right]}{\alpha\eta(\mu)}\right\},\quad G_2(t,\mu)=\frac{2e^{-\frac{1}{2}\alpha^2Dt}}{\Delta(\mu)}\left\{\frac{\sinh\left[\frac{1}{2}\alpha t\eta(\mu)\right]}{\alpha\eta(\mu)}\right\}. \tag{10}$$

Note that functions (10) satisfy equation (4) and following conditions

$$G_1(0,\mu)=1,\quad G_1(\tau,\mu)=0,\quad G_2(0,\mu)=0,\quad G_2(\tau,\mu)=1.$$

IOP Conf. Series: Materials Science and Engineering 1016 (2021) 012015 doi:10.1088/1757-899X/1016/1/012015

In formula (9) the expressions $\phi_1\left(\frac{\partial}{\partial\mu}\right)$, $\phi_2\left(\frac{\partial}{\partial\mu}\right)$ is obtained from $\phi_1(x)$, $\phi_2(x)$ replacing x by differentiatial symbol $\frac{\partial}{\partial\mu}$. For each one-term quasipolynomial (8), the corresponding differential quasipolynomial $P\left(\frac{\partial}{\partial\mu}\right)e^{\beta\frac{\partial}{\partial\mu}}$ acts on the functions in curly braces of formula (9) by the formula

$$P\left(\frac{\partial}{\partial\mu}\right)e^{\beta\frac{\partial}{\partial\mu}}\left\{G(t,\mu)e^{\mu x}\right\}\Big|_{\mu=0}=P\left(\frac{\partial}{\partial\mu}\right)\left\{G(t,\mu)e^{\mu x}\right\}\Big|_{\mu=\beta}.$$

5. Examples of solving problems for specific parameters of telegraph equation with known behavior of process at two moments of time

Let's consider the examples of the process, which is described by problem (2), (3) and is determined by the specific parameters α, D, C with given behavior of the process at two points in time.

Example 1. We study problem (2), (3) for the case $\phi_1(x)=c_1$, $\phi_2(x)=c_2$, $\tau=1$, where c_1,c_2 are positive constants. Since the functions $\phi_1(x)=c_1e^{0x}$, $\phi_2(x)=c_2e^{0x}$ as one-term quasipolynomials of form (8) have index 0 in exponential functions, then $\eta(0)=\left(D^2\alpha^2-4E\right)^{1/2}$.

We consider three cases.

1) If $D^2\alpha^2-4E>0$, then $\alpha^2\left(D^2\alpha^2-4E\right)\tau^2\neq-4\pi^2k^2$ for each $k\in\mathbb{N}$. It means that $0\notin L$. Let's write functions (10) for $\mu=0$:

$$G_1(t,0)=e^{-\frac{1}{2}\alpha^2Dt}\left\{\frac{\sinh\left[\frac{1}{2}\alpha(1-t)\eta(0)\right]}{\sinh\left[\frac{\alpha}{2}\eta(0)\right]}\right\},\quad G_2(t,0)=e^{-\frac{1}{2}\alpha^2D(t-1)}\left\{\frac{\sinh\left[\frac{1}{2}\alpha t\eta(0)\right]}{\sinh\left[\frac{\alpha}{2}\eta(0)\right]}\right\}.$$

The solution of problem (2), (3) by formula (9) has following form

$$u(t,x)=c_1G_1(t,0)+c_2G_2(t,0)$$
$$=\frac{e^{-\frac{1}{2}\alpha^2Dt}}{\sinh\left[\frac{\alpha}{2}\eta(0)\right]}\left(c_1\sinh\left[\frac{1}{2}\alpha(1-t)\eta(0)\right]+c_2e^{\frac{1}{2}\alpha^2D}\sinh\left[\frac{1}{2}\alpha t\eta(0)\right]\right).$$

For example, for $\alpha=1$, $D=3$, $E=2$ we get solution of problem (2), (3) in analytical form

$$U(t,x)=\frac{e^{-2t}}{e-1}\left(c_1\left(e-e^t\right)+c_2e^2\left(e^t-1\right)\right).$$

Note that the function $u(t,x)$ at each point x is the same, i.e. the behavior of process does not depend on the coordinate x and it goes exponentially to zero, if $t\to\infty$.

Graphic representation of the solution is shown in Figure 1a), 1b) and 1c).

IOP Conf. Series: Materials Science and Engineering 1016 (2021) 012015 doi:10.1088/1757-899X/1016/1/012015

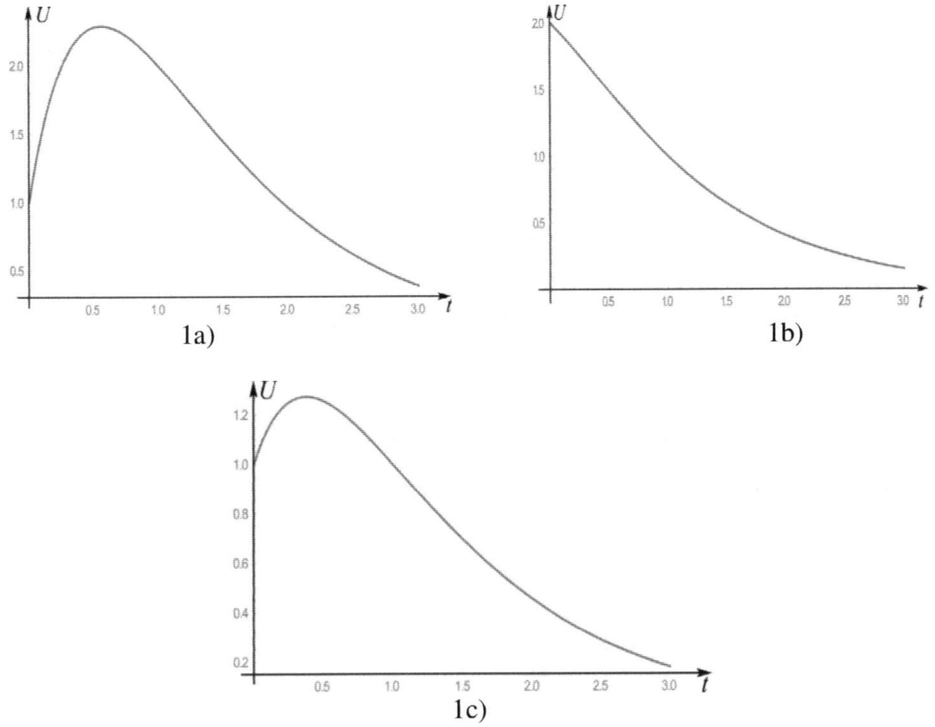

1a) 1b)

1c)

Figure 1. The graph of the function $u(t,x)$
for a) $c_1 = 1,\ c_2 = 2$, b) $c_1 = 2,\ c_2 = 1$ та c) $c_1 = c_2 = 1$

2) If $D^2\alpha^2 - 4E = 0$, then $\eta(0) = 0$ and the condition $0 \notin L$ is fulfilled. This expression is satisfied, for example, for $E = 0$ and $D = 0$, i.e. the conductor is well insulated and the resistance is insignificant. Functions (10) for $\mu = 0$ have the such form

$$G_1(t,0) = (1-t)e^{-\frac{1}{2}\alpha^2 Dt}, \quad G_2(t,0) = e^{-\frac{1}{2}\alpha^2 D(t-1)}t.$$

We find solution of problem (2), (3) by formula (9):

$$U(t,x) = \left(c_1(1-t) + c_2 e^{\frac{1}{2}\alpha^2 D} t \right) e^{-\frac{1}{2}\alpha^2 Dt}.$$

In particular, for $\alpha = 1,\ D = 2,\ E = 1$ we get solution of problem in analytical form

$$U(t,x) = c_1(1-t)e^{-t} + c_2 t e^{1-t}$$

and graphic form (Figure 2)

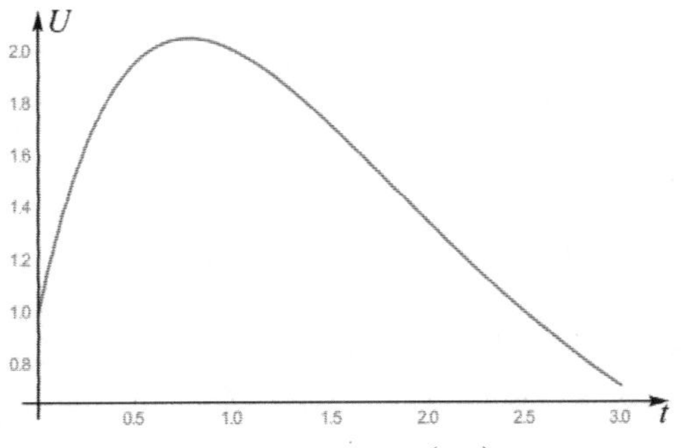

Figure.2. Graph of the function $u(t,x)$ for $c_1 = 1,\ c_2 = 2$.

3) Let $D^2\alpha^2-4E<0$. Then provided that value $\frac{\alpha^2\left(4E-D^2\alpha^2\right)\tau^2}{\pi^2}$ is not a multiple of 4, we get $0\notin L$. Functions (10) for $\mu=0$ have following form

$$G_1(t,0)=e^{-\frac{1}{2}\alpha^2 Dt}\left\{\frac{\sin\left[\frac{1}{2}\alpha(1-t)\sqrt{4E-D^2\alpha^2}\right]}{\sin\left[\frac{\alpha}{2}\sqrt{4E-D^2\alpha^2}\right]}\right\},\quad G_2(t,0)=e^{-\frac{1}{2}\alpha^2 D(t-1)}\left\{\frac{\sin\left[\frac{1}{2}\alpha t\sqrt{4E-D^2\alpha^2}\right]}{\sin\left[\frac{\alpha}{2}\sqrt{4E-D^2\alpha^2}\right]}\right\}.$$

By formula (9), the solution of problem (2), (3) has the form

$$U(t,x)=c_1G_1(t,0)+c_2G_2(t,0)$$
$$=\frac{e^{-\frac{1}{2}\alpha^2 Dt}}{\sin\left[\frac{\alpha}{2}\sqrt{4E-D^2\alpha^2}\right]}\left(c_1\sin\left[\frac{1}{2}\alpha(1-t)\sqrt{4E-D^2\alpha^2}\right]+c_2e^{\frac{1}{2}\alpha^2 D}\sin\left[\frac{1}{2}\alpha t\sqrt{4E-D^2\alpha^2}\right]\right).$$

In particular, for $\alpha=1,\ D=2,\ E=2$ we get the solution of problem (2), (3) in the form

$$U(t,x)=\frac{e^{-t}}{\sin 1}\left(c_1\sin[1-t]+c_2e\sin[t]\right).$$

A graphical representation of this solution is given in Figure 3.

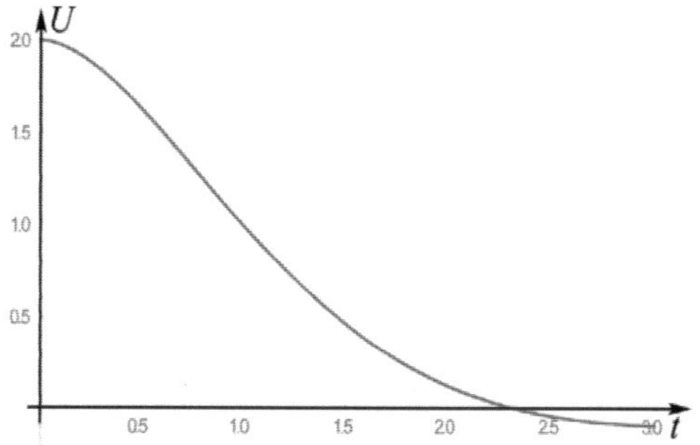

Figure. 3. The graph of the function $u(t,x)$ for $c_1=2,\ c_2=1$.

Example 2. Let's consider process in electrical engineering system which is modeled by two-point problem

$$\left[-\frac{\partial^2}{\partial x^2}+\frac{\partial^2}{\partial t^2}+4\frac{\partial}{\partial t}+3\right]u(t,x)=0, \tag{11}$$

$$u(0,x)=\sin x,\ u(1,x)=0. \tag{12}$$

Equation (11) is telegraph equation (2) for $\alpha=1,\ D=4,\ E=3$. Two-point conditions (12) are conditions (3) for $\tau=1,\ \phi_1(x)=\sin x,\ \phi_2(x)=0$.

The function $\phi_1(x)=\sin x=\frac{1}{2i}\left(e^{ix}-e^{-ix}\right)$ is the sum of two one-term quasipolynomials with coefficients i and $-i$ in exponential functions. Since $\eta(\pm i)=0$ and $\Delta(\pm i)=e^{-2}\neq 0$ then function $\phi_1(x)$ is a quasipolynomial from the class $K_{\mathbb{C}\setminus L}$. According to formula (9) the solution of problem (11), (12) has the form

$$U(t,x)=\frac{1}{2i}\left(\left\{G_1(t,\mu)e^{\mu x}\right\}\Big|_{\mu=i}-\left\{G_1(t,\mu)e^{\mu x}\right\}\Big|_{\mu=-i}\right)$$

$$=\frac{1}{2i}\left(\left\{\frac{2e^{-2(t+1)+\mu x}}{\Delta(\mu)}\left\{\frac{\sinh\left[\frac{1}{2}(1-t)\eta(\mu)\right]}{\eta(\mu)}\right\}\right\}\Bigg|_{\mu=i}-\left\{\frac{2e^{-2(t+1)+\mu x}}{\Delta(\mu)}\left\{\frac{\sinh\left[\frac{1}{2}(1-t)\eta(\mu)\right]}{\eta(\mu)}\right\}\right\}\Bigg|_{\mu=-i}\right).$$

$$=\frac{(1-t)}{2i}\left(e^{-2t+\mu x}\Big|_{\mu=i}-e^{-2t+\mu x}\Big|_{\mu=-i}\right)=(1-t)e^{-2t}\sin x.$$

The graph of this function is represented in Figure 4.

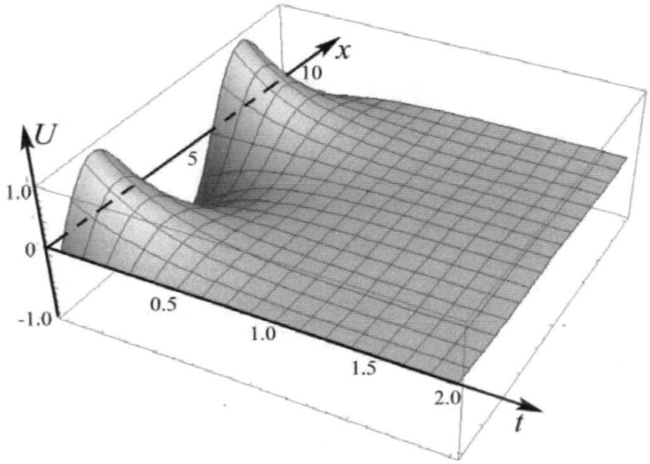

Figure 4. The graph of the function $U(t,x)$

Note that the obtained solution of problem (11), (12) is 2π-periodical function by variable x and for $t\to\infty$ it goes to zero.

Conclusions

The process of electric current distribution in an infinitely long conductor with given system characteristics and process behavior at two moments in time is described by a two-point problem for the telegraph equation. A class of quasipolynomials in which there is a unique solution of the problem is distinguished. According to the differential-symbol method, the formula for finding solution is proposed. The obtained solution of the problem makes possible to determine the process of electric current propagation in the conductor at any time and at any point of the conductor, which greatly facilitates the design of electrical systems. The application of the method is shown in the examples.

References

[1] Sinha R, Paredis Ch, Liang V and Khosla P 2001 Modeling and simulation methods for design of engineering systems *J. Comp. Inf. Sci. Eng* **1** 84-91
[2] Vepryk Y, Nebera O 2015 Modeling of power supply systems with motor load in transient modes *Technological audit and production reserves* **6** 55-59
[3] Vepryk Y and Lebedka S 2012 Mathematical modeling of modes of operation of electric networks with OPN *Eastern-Europ. J. of Enterprise Technologies* **8** 25-29
[4] Manchanda P, Lozi R and Hasan A 2020 *Mathematical modelling, optimization, analytic and numerical solution* (Singapore: Springer)

[5] Magrab E B 2012 *Vibrations of Elastic Systems with Applications to MEMS and NEMS* (New York: Springer)
[6] Velten K 2009 *Mathematical modeling and simulation: introduction for scientists and engineers* (Wiley-VCH) p 348
[7] Dulay M, Oltean S-E and Duka A.-V 2016 Modeling and simulation of the operation of a mechanical system which is affected by uncertainties. *Proc. Technology. 9th International Conference Interdisciplinarity in Engineering, INTER-ENG* **22** 662-669
[8] Jordan P M and Puri A 1999 Digital signal propagation in dispersive media *J. Appl. Phys* **85** 1273-82
[9] Huo X -Y and Lei Y-Z 2010 A new method for calculating transient electromagnetic responses of AC/DC power system with external electromagnetic pulse interference *Progress in Electromagnetics Research* **2010** 245-260
[10] Chaban A, Levoniuk V and Lysiak V 2017 Analysis of transitional processes in the long line of electric transmissions as an element of the electric power system *Bulletin of Vinnytsia Polytechnic Institute* **2** 63-69
[11] Stakhiv P, Kozak Y and Hoholyuk O 2013 Discrete mathematical macromodel of electric transmission line *Przeglad elektrotechniczny* **4** 272-274
[12] Strelkovska V and Paskalenko V 2017 *The equation of mathematical physics for specialists in the field of communication* (Odessa: Popova National Academy of Communication)
[13] Jordan P M and Puri A 1999 Digital signal propagation in dispersive media *J. Appl. Phys* **85** 1273–1282
[14] Goldstein S 1951 On diffusion by discontinuous movements and the telegraph equation *Quart. J. Mech. Appl. Math.* **4** 129–156
[15] Banasiak J, and J. R. Mika 1998 Singularly perturbed telegraph equations with applications in the random walk theory *J. Appl. Math. Stoch. Anal.* **11** 9-28
[16] Weston V H and He S 1993 Wave splitting of the telegraph equation in R3 and its application to inverse scattering *Inverse Probl.* **9** 789–812
[17] Nitrebich Z 1996 A boundary-value problem in an unbounded strip *J. Math. Sci.* **79** 1388-1392
[18] Malanchuk O and Nytrebych Z 2017 Homogeneous two-point problem for PDE of the second order in time variable and infinite order in spatial variables *Open Math.* **15** 101–110
[19] Nytrebych Z and Malanchuk O 2019 The conditions of existence of a solution of the degenerate two-point in time problem for PDE *Asian-European Journal of Mathematics* **12** 1950037
[20] Nytrebych Z and Malanchuk O 2019 The conditions of existence of a solution of the two-point in time problem for nonhomogeneous PDE *Italian J. of Pure and Appl. Mathematics* **41** 242–250
[21] Nytrebych Z and Malanchuk O 2017 The differential-symbol method of solving the problem two-point in time for a nonhomogeneous partial differential equation *J. Math. Sci.* **227** 68–80
[22] Ostapenko V 2008 The first boundary value problem for the telegraph equation in a semibounded domain *Diff. equations and their applications DNU* **2008** 18-20
[23] Ostapenko V 2009 Neumann boundary value problem for telegraph equation in the semibounded domain *Bulletin DNU. Series Simulation* **8**, 89–92
[24] Ostapenko V 2012 Mixed initial-boundary value problem for telegraph equation in domain with variable borders *Advanced in Mathematical Physics* **5** 831012
[25] Nytrebych Z, Ilkiv V, Pukach P, Malanchuk O, Kohut I and Senyk A 2019 Analytical method to study a mathematical model of wave processes under two-point time conditions *Eastern-Europ. J. of Enterprise Technologies* **1** 74–83
[26] Nytrebych Z and Malanchuk O 2019 The differential-symbol method of constructing the quasipolynomial solutions of two-point problem *Demonstratio Mathematica* **52** 88-96
[27] Nytrebych Z and Malanchuk O 2017 The differential-symbol method of solving the two-point problem with respect to time for a partial differential equation *J. Math. Sci.* **224** 541–554

IOP Conf. Series: Materials Science and Engineering 1016 (2021) 012016 doi:10.1088/1757-899X/1016/1/012016

Information Technology of Robotic Prosthesis Computer-Aided Design Based on Parametric Modeling

A V Parkhomenko*, O M Gladkova, Y I Zalyubovskiy and A V Parkhomenko

National University Zaporizhzhia Polytechnic, 64, Zhukovskogo str., Zaporizhzhia, 69063, Ukraine

*Email: parhom@zntu.edu.ua

Abstract. The features of modern technologies of three-dimensional parametric modeling were investigated in this paper. It was shown that the application of tabular parameterization makes it possible to increase the efficiency of the computer-aided design of biomedical products by accelerating the development of similar components in their composition. An information technology for the computer-aided design of a robotic prosthesis prototype in the PTC Creo environment has been developed. The functionality of the virtual and physical prototypes of the robotic prosthesis was checked with the usage of digital kinematic simulation and real tests. The developed tabular parametric model makes it possible to control the mass-dimensional characteristics of the product during the development of individual prostheses, taking into account the anatomical features of a particular patient.

1. Introduction

The development of modern technologies in the field of bioengineering provides new opportunities, ranging from smart implants and prostheses to 3D printing of tissues and even organs. Earlier a person, who having lost a limb, felt significant discomfort and unnaturalness while wearing a prosthesis. Then modern prostheses have become more approximate both functionally and visually. For example, in work [1], a whole set of hand prostheses was presented, and the same testing procedure had been performed for all of them.

The investigations have shown that the most important parameters influencing the quality indicators of the prosthesis are the hand weight [2-3], the size and anatomically correct kinematics of fingers, the type of drive and the drive mechanism, the force and the speed of grip, durability, and appearance [4]. At the same time, a compromise must be found between anatomical accuracy and reliability, weight, complexity and cost of the product. Thus, there are many problems associated with the design of robotic prostheses, taking into account the individual characteristics of patients, the level of needs and activity. The complexity of parallel design of mechanical and electronic parts of the prosthesis is one of them [5]. The parameters and characteristics of the real product can be improved, and its cost can be reduced based on modern technologies of computer-aided design (CAD) and investigation of virtual prototype of the robotic prosthesis [6-7].

The goal of this work is the development of information technology of computer-aided design of a virtual prototype of the mechanical part of the robotic prosthesis based on parametric modeling.

2. Analysis of the features of a parametric modelling

Parametric modeling is used today in various fields of design, in particular in mechanical engineering and architecture [8]. However, its application in the development of prostheses and implants requires consideration of the specifics of the designed products and verification of the results based on comprehensive study of 3D virtual and physical prototypes.

The process of parametric modeling is based on the usage of model elements parameters and interrelations between them. Parameterization gives the opportunity to go through various design schemes in a short time (by changing the parameters or geometric relations) and avoid fundamental errors. Parametric modeling differs significantly from conventional two-dimensional drawing or three-dimensional modeling. In the case of parametric design, the developer creates a mathematical model of objects with parameters. Their change leads to changes in the configuration of the part and the mutual movement of the parts in the assembly [9].

Parameterization of 2D drawings is usually available in CAD-systems of medium and lightweight. However, the emphasis in these systems is on 3D design technology, and the possibility of two-dimensional drawings parameterizing is practically not used. Parametric CAD systems focused on two-dimensional drawings are often "truncated" versions of more advanced CAD. Three-dimensional parametric modeling is a much more efficient (but also more complex) tool than two-dimensional parametric modeling. In modern systems of medium and heavyweight, the presence of a parametric model is embedded in the ideology of CAD itself, and the parametric description of the object is the basis for the entire design process.

As studies have shown, there are different types of parameterization (Fig. 1), which have their own features, advantages and disadvantages.

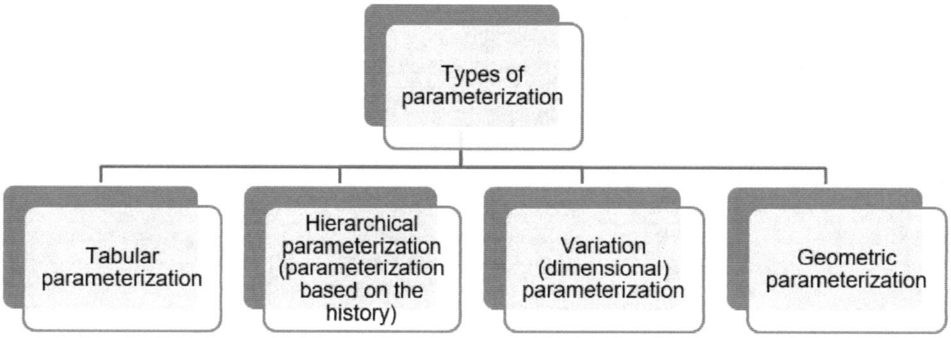

Figure 1. Classification of parameterization types

Tabular parameterization is based on the creation of a table of parameters for typical parts. The development of a new instance of the part is carried out by selection from the table of standard sizes. The possibilities of tabular parameterization are limited because it is usually impossible to set arbitrary new values of parameters and geometric relations. However, tabular parametrization is widely used in all parametric CAD-systems as it can significantly simplify and speed up the creation of libraries of standard and typical parts, as well as their use in the design process [9]. That is why tabular parameterization can be recommended for use in the automated design of prostheses and implants to take into account various standard sizes, depending on the individual anatomical characteristics of patients.

3. Information technology of parametric modelling

According to research [10], PTC Creo is one of eight the best parametric modeling software in the world. The system has powerful functionality and provides an intelligent design environment. That is why this system was chosen for parametric modeling of a robotic hand prosthesis. The main stages of the developed information technology are the following.

1. Creation a structural diagram of a designed product based on the principles of hierarchy and decomposition (Fig. 2).

2. Analysis of the design features of the components of the product and identification of features (notches, channels, etc.) to ensure kinematic properties.

3. Design of individual parts (fingers) using tabular parameterization. Table 1 contains the following parameters: F_HEIGHT - finger height; F_WIDTH - finger width; F_LENGHT - finger length; F_F_HEIGHT - height of the main part of the finger; F_WIDTH_END - width of the terminal phalanx; SMALL_F - length of the terminal phalanx of the finger; F - length of the main phalanx of the finger.

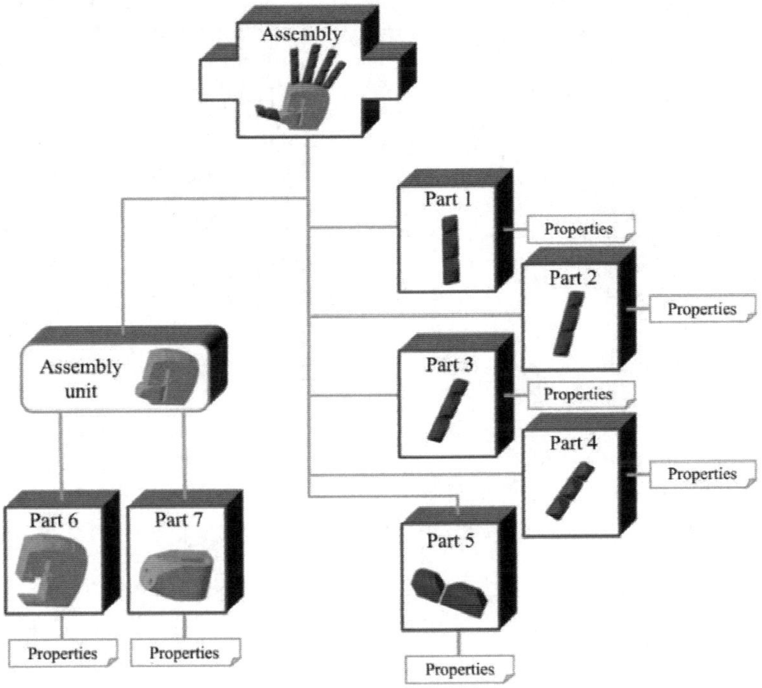

Figure 2. Structural diagram of a designed product

The parameters F_HEIGHT, F_WIDTH, F_LENGHT are entered by the designer independently, based on the dimensions of the required finger. The parameters F_F_HEIGHT, F_WIDTH_END, SMALL_F, F are calculated using simplified expressions that were obtained based on the anatomy of an average person's finger: *F_F_HEIGHT = F_HEIGHT – 9; F_WIDTH_END = F_WIDTH – 1mm; SMALL_F = (F_HEIGHT / 3) * 4; F = (F_LENGHT - (SMALL_F + 2 mm)) / 2*, where 2 mm is the sum of the distances between three phalanges and 2 is the number of phalanges.

4. Building and rendering of the final assembly (Fig. 3a). Checking the functionality of the product with digital kinematic simulation tools.

5. Generation of files for printing the components of the product on a 3D printer.

6. 3D printing, assembling and testing the physical prototype (Fig.3b).

Table 1. 3D parametric model

Parameter	Index finger, mm	Middle finger, mm	Ring finger, mm	Small finger, mm	Thumb, mm
F_HEIGHT	13,2	14,1	13,2	12	16
F_WIDTH	14	15	14	13	18
F_LENGHT	71,6	77,8	71,6	56,7	80
F_F_HEIGHT	4,2	5,1	4,2	3	7
F_WIDTH_END	13	14	13	12	17
SMALL_F	17,6	18,8	17,6	16	21,333
F	26	28,5	26	19,35	28,333

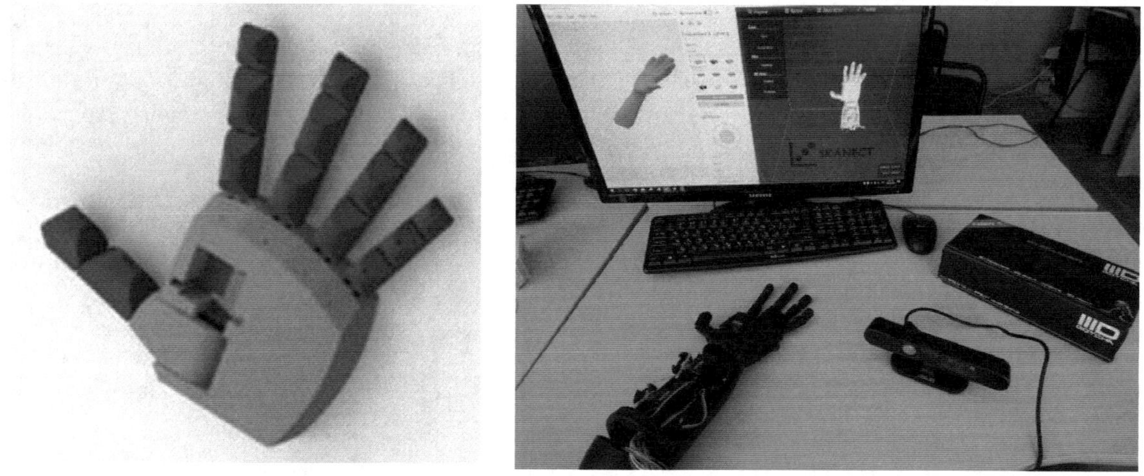

a b

Figure 3. Assembling (a) and testing (b) the prototype of the robotic prosthesis

Conclusions

The usage of prototyping technologies in computer-aided design of robotic prostheses opens ways to further improvement of their shape, weight and size characteristics and control algorithms. Creation and research of a virtual prototype of a designed product allows to identify errors at the early stages of project creation and, accordingly, reduce the development time and costs for revision; to explore more alternatives and ways of project realization; to perform prototype investigations, which are sometimes impossible in real conditions on a full-scale model; to perform fewer physical experiments that require significant time and material costs.

References

[1] Belter T J Segil L J Aaron M D and Weir F R 2013 Mechanical design and performance specifications of anthropomorphic prosthetic hands: A review *J. of Rehabilitation Research & Development* vol **50** (5) pp 599-618

[2] Pons J L Rocon E Ceres R Reynaerts D Saro B Levin S and W Van Moorleghem 2004 The MANUS-HAND dextrous robotics upper limb prosthesis: mechanical and manipulation aspects *Autonomous Robots* vol **16** pp 143–163

[3] Light C M and Chappell P H 2000 Development of a lightweight and adaptable multiple-axis hand prosthesis *Medical Engineering & Physics* vol **22** (10) pp 679-684

[4] Zheng J Z De La Rosa S and Dollar A M 2011 An investigation of grasp type and frequency in daily household and machine shop tasks Proceedings of the IEEE International Conference on Robotics and Automation (ICRA) (Shanghai, China) pp. 4169–4175

[5] Parkhomenko A Gladkova O and Zalyubovskiy Y 2019 *Investigation and realization of prototyping technologies for robotic-prostheses computer aided design* Proceedings of the XV International Conference on The Experience of Designing and Application of CAD Systems in Microelectronics (CADSM 2019) (Lviv: Lviv Polytechnic) pp 7/5-7/8

[6] Liou W F 2007 *Rapid Prototyping and Engineering Applications: A Toolbox for Prototype Development* (CRC Press Taylor & Francis Group Boca Raton) 592p

[7] McEwen A and Cassimally H 2014 *Designing the Internet of things* (Wiley) 338 p.

[8] Kalkan E, Okur F Y and Altunisik A C 2018 Applications and usability of parametric modeling *Journal of Construction Engineering, Management & Innovation* vol **1** (3) pp 139-146

[9] Malykh V N 2010 *Introduction to modern CAD* (Moscow: DMK Press, in Russian) 192 p.

[10] Gaget L 2018 Top 8 of the best parametric modeling software in 2020 (https://www.sculpteo.com/blog/2018/03/07/top-8-of-the-best-parametric-modeling-software/)

The Use of Cellular Automata in the Study of Heat and Mass Transfer Processes in Particular During Wood Drying

Ya Sokolovskyy*, O Sinkevych, M Levkovych, V Shymanskyi and V Yarkun

IT Department, Ukrainian National Forestry University, Lviv, UKRAINE

*Email: sokolowskyyyar@yahoo.com

Abstract. This paper demonstrates the use of cellular automata in the study of heat and mass transfer processes, in particular during wood drying. For study, we developed a three-dimensional mathematical model. This model is described by a system of interconnected differential equations. These equations are described in partial time derivatives and coordinates in space. The peculiarity of this work is the possibility of combining a three-dimensional mathematical model and the theory of cellular automata in solving the problem. Usage of the theory of cellular automata requires the presence of a certain number of identical elements of the system. In this case, we propose to present the studied CAD model in the form of cubes of the same size. The paper also describes the relationships between the created cubes. In addition, we give the algorithm of calculations. Based on the described algorithm, we carried out the number of experiments, which prove the possibility of using the theory of cellular automata in solving the problem.

1. Introduction

As practice shows, when we drying wood, the various physical processes and phase transitions occur. These processes directly affect the change of mechanical and physical properties of the material, its geometric shape, presence of cracks and the possibility of collapse. In this regard, the task of creating new and increasingly effective methods of wood drying control remains very important [1].

When we are modeling such systems, we often encountered complex initial and boundary conditions. This makes it difficult to obtain reliable analytical solutions when we solve such problems. Because of this, we often use a various numerical methods, which are often not stable. In turn, if we use differential equations in the modeling of physical processes it does not always allow us to obtain an acceptable result. In this regard, there is an important need to develop effective software and mathematics with the possibility of their implementation with algorithmic tools [2].

In performing this work, we use the finite element method for the existing three-dimensional mathematical model. However, the use of this method requires significant time, especially at very high densities. In this regard, we propose to use the theory of cellular automata, which should significantly speed up the calculation process even at very high densities.

One of the most important contributions to the development of the theory of cellular automata made by S. Wolfram, who in his work [3] gives their detailed description and possibilities of application for almost all branches of science. At the same time, in [4] we can see the rules of transitions when using cellular automata, taking into account their internal variables. This paper also presents the possibilities of using cellular automata in modeling various heat transfer processes and spatial dynamics.

In works [5-9] we can see the features of the use of composite materials by using effective mechanical properties based on micro level cellular structural models. In this regard, we can assume that we can use cellular automata in modeling different dynamic systems that have a characteristic similarity between the basic elements. All this allows us to outline the prospects of this area.

2. Description of the studied model and its presentation

In this work, we have 3D model of wood. To determine the value of temperature and humidity in the middle of this model, we use a three-dimensional mathematical model of heat and mass transfer and the theory of cellular automata [10].

First, we need to present 3D model of wood in the form of a set of cubes. We can do that by considering the physical size of the 3D model of wood and the density of its separation. This representation allows us to use the theory of cellular automata [11, 12].

Next, we need to define the boundary cubes. We consider a boundary cube if the number of his adjacent cubes is less than six. To determine the number of adjacent cubes it is necessary to know their location. To do this, we used the neighborhoods of von Neumann of the first order (see Figure 1).

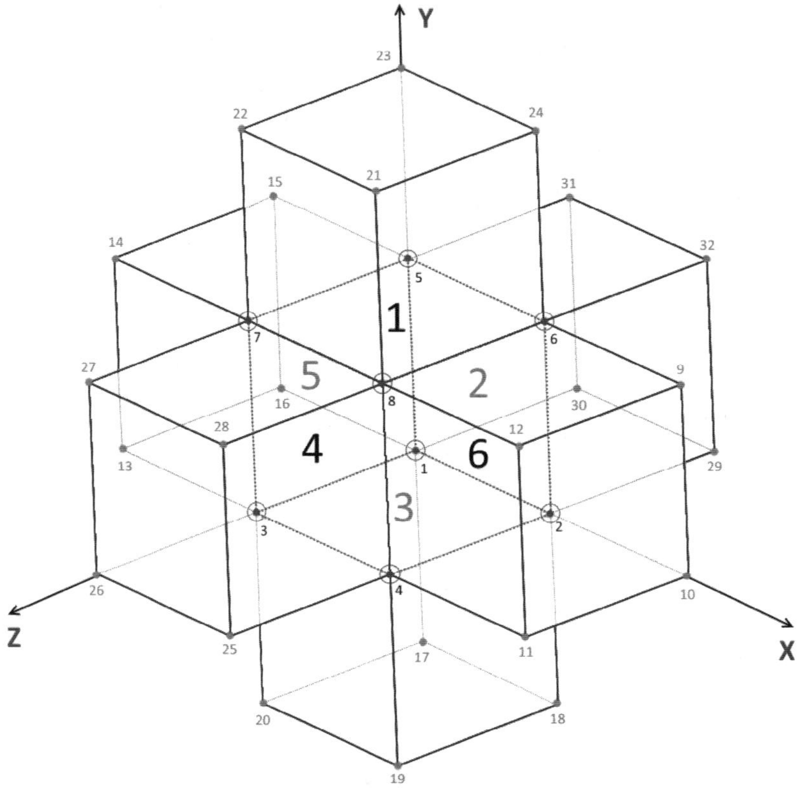

Figure 1. Location of cubes in space according to the neighborhood of von Neumann.

Cubes that have six adjacent cubes are internal. Each cube in our system has eight main points, which are located at the corners of the cube. One of these points is the target. The coordinates of this point correspond to the location of the cube according to the grid. Knowing the coordinates of the target point, we can determine the coordinates of the other seven main points [13].

Having the coordinates of the eight main points of the cube, we can determine the coordinates of the tangents of adjacent cubes by using the binary code, which you can see in table. 1.

Table 1. The coordinates of the tangents of adjacent cubes in the binary code.

1	2	3	4	5	6	7	8
0	1	0	1	0	1	0	1
0	0	0	0	1	1	1	1
0	0	1	1	0	0	1	1

Depending on the selected direction of displacement (x + 1, x – 1, y + 1, y – 1, z + 1, z – 1), the values of 0 and 1 will be represented by certain coordinates, which are given in table 2, and the values x, y, z will correspond to the coordinates of the target point of the main cube.

Table 2. The values of the coordinates relative to the selected direction of displacement.

	0	**1**
X + 1	X + 1	X + 2
	Y	Y + 1
	Z	Z + 1
X - 1	X - 1	X
	Y	Y + 1
	Z	Z + 1
Y + 1	X	X + 1
	Y + 1	Y + 2
	Z	Z + 1

	0	**1**
Y - 1	X	X + 1
	Y - 1	Y
	Z	Z + 1
Z + 1	X	X + 1
	Y	Y + 1
	Z + 1	Z + 2
Z - 1	X	X + 1
	Y	Y + 1
	Z - 1	Z

3. Description of the mathematical model

In this work, we use a three-dimensional mathematical model of heat transfer of wood during its drying [14]. The description of this mathematical model by means of system of differential equations in partial derivatives on spatial coordinates and time has the following form:

$$\begin{cases} c\rho \dfrac{\partial T^{(i)}}{\partial t} = \sum\limits_{j=1}^{3} \lambda_j \dfrac{\partial^2 T^{(i)}}{\partial x_j^2} + \varepsilon \rho_0 r \dfrac{\partial U^{(i)}}{\partial t} \\ \dfrac{\partial U^{(i)}}{\partial t} = \sum\limits_{j=1}^{3} a_j \dfrac{\partial^2 U^{(i)}}{\partial x_j^{\,2}} + \delta \sum\limits_{j=1}^{3} a_j \dfrac{\partial^2 T^{(i)}}{\partial x_j^{\,2}} \end{cases} \tag{1}$$

where ρ is density, x, y, z are represent the coordinates, i is serial number of cube, $T^{(i)}$ (t, x_1, x_2, x_3) is the temperature of the i-th cube, t is simulation time and c is specific heat. In turn, ρ_0 is basic density, a_j are humidity coefficients, λ_j are the coefficients of thermal conductivity, $U^{(i)}$ (t, x_1, x_2, x_3) is the moisture content of the i-th cube, r is specific heat of vaporization, δ is the thermo gradient coefficient and ε is phase transition coefficient. After that, we introduce a time-space grid in area $G = \{(t, x_1, x_2, x_3)\text{: } 0 \le t \le \tau,\ 0 \le x_1, x_2, x_3 \le l\}$:

$$\begin{aligned} &\Omega_{\Delta t, h_j (j=\overline{1,3})} = \{(t^k, x_{1(n)}, x_{2(m)}, x_{3(p)}) : t^k = k\Delta t, k = \overline{0,K}, \\ &\Delta t = \frac{\tau}{K}; x_{1(n)} = (n-1)h_1, n = \overline{1,N}, h_1 = \frac{l}{N-1};\ x_{2(m)} = (m-1)h_2, \\ &m = \overline{1,M}, h_2 = \frac{l}{M-1};\ x_{3(p)} = (p-1)h_3, p = \overline{1,P}, h_3 = \frac{1}{P-1};\} \end{aligned} \tag{2}$$

The initial conditions for a constant rate of wood drying has the following form:

$$T^{(i)}\Big|_{t=0} = T_0^{(i)} \qquad U^{(i)}\Big|_{t=0} = U_0^{(i)} \tag{3}$$

For the period of falling speed of the drying process of capillary-porous materials, the initial moisture content can be describe by the following quadratic function:

$$\begin{aligned} &U^{(i)}\Big|_{t=0} = U_c^{(i)} - \left[1 - \left(\frac{x_1 - l_1/2}{l_1/2}\right)^2\right] \times \left[1 - \left(\frac{x_2 - l_2/2}{l_2/2}\right)^2\right] \times \\ &\times \left[1 - \left(\frac{x_3 - l_3/2}{l_3/2}\right)^2\right] \times (U_c - U_s) \end{aligned} \tag{4}$$

where U_c is the moisture content in the center and U_s is the moisture content on the surface.

For a three-dimensional mathematical model, we used boundary conditions of the third kind [15], each of which corresponds to one coordinate. The boundary conditions for the X coordinate have the following form:

$$\begin{cases} \lambda_1 \dfrac{T^{k(i)}_{N,m,p} - T^{k(i)}_{N-1,m,p}}{h_1} + \rho_0(1-\varepsilon)\beta(U^{k(i)}_{N,m,p} - U_p) = \alpha(T^{k(i)}_{N,m,p} - T_c) \\ a_1\delta \dfrac{T^{k(i)}_{N,m,p} - T^{k(i)}_{N-1,m,p}}{h_1} + a_1 \dfrac{U^{k(i)}_{N,m,p} - U^{k(i)}_{N-1,m,p}}{h_1} = \beta(U_p - U^{k(i)}_{N,m,p}) \end{cases} \quad (5)$$

where β is the moisture exchange coefficient, α is the coefficient of heat exchange, l is the length of one face of cube. $U_p = F(T_c, \varphi)$, where T_c is a temperature and φ is a relative humidity of the drying agent. The three-dimensional mathematical model has the following finite-difference form:

$$c\rho \frac{T^{k+1(i)}_{n,m,p} - T^{k(i)}_{n,m,p}}{\Delta t} = \sum_{j=1}^{3} \frac{\lambda_j}{h_j^2}\left(T^{k(i)}_{\omega_j+1,\gamma_j} - 2T^{k(i)}_{\omega_j,\gamma_j} + T^{k(i)}_{\omega_{j-1},\gamma_j}\right) + \varepsilon\rho_0 r \frac{U^{k+1(i)}_{n,m,p} - U^{k(i)}_{n,m,p}}{\Delta t} \quad (6)$$

$$\frac{U^{k+1(i)}_{n,m,p} - U^{k(i)}_{n,m,p}}{\Delta t} = \sum_{j=1}^{3} \frac{a_j}{h_j^2}\left(U^{k(i)}_{\omega_{j+1},\gamma_j} - 2U^{k(i)}_{\omega_j,\gamma_j} + U^{k(i)}_{\omega_j-1,\gamma_j}\right) + $$
$$+ \delta \sum_{j=1}^{3} a_j \left(T^{k(i)}_{\omega_{j+1},\gamma_j} - 2T^{k(i)}_{\omega_j,\gamma_j} + T^{k(i)}_{\omega_{j-1},\gamma_j}\right) \quad (7)$$

where $\omega_1 = n,\ \gamma_1 = m, p;\ \omega_2 = m,\ \gamma_2 = n, p;\ \omega_3 = p,\ \gamma_3 = n, m;$

4. Description of cell-automatic interaction

When we use cellular automata, we often use the coefficients of moisture and thermal conductivity. These coefficients used when we transmitting numerical values of temperature and humidity within one cube and provided that it has other points in addition to the main 8 points. Other points will exist if our step on the coordinates h has a value that is less than one. Therefore, we have dependences by which we can transfer numerical values between different points of one cube according to the scheme of relations within the selected edge [16].

$$H^{(i)}[1,3] = \frac{K_2 H^{(i)}[1,3] + C_1^{(i)}}{K_2 + C_1} \qquad H^{(i)}[5,6] = \frac{K_1 H^{(i)}[5,6] + C_2^{(i)}}{K_1 + C_2} \qquad H^{(i)}[2,4] = \frac{K_3 H^{(i)}[2,4] + C_3^{(i)}}{K_3 + C_3} \quad (8)$$

where if $H^{(i)} = T^{(i)}$ then $K = \lambda$ and if $H^{(i)} = U^{(i)}$ then $K = a$. In addition:

$$C_1^{(i)} = \begin{cases} K_1 H^{(i)}[5,6], C_1 = K_1 \\ K_3 H^{(i)}[2,4], C_1 = K_3 \end{cases} \qquad C_2^{(i)} = \begin{cases} K_2 H^{(i)}[1,3], C_2 = K_2 \\ K_3 H^{(i)}[2,4], C_2 = K_3 \end{cases} \qquad C_3^{(i)} = \begin{cases} K_1 H^{(i)}[5,6], C_3 = K_1 \\ K_2 H^{(i)}[1,3], C_3 = K_2 \end{cases}$$

We also use cell-automatic interaction between two points of adjacent cubes. Each of these points has the values of temperature (T_1, T_2) and humidity (U_1, U_2). Therefore, if there is a temperature interaction between two points, then we follow the following rules:

$$T_k = T_1 + (T_2 - 0.5 \cdot (T_1 + T_2)) \quad (9)$$

if $T_k > T_c$ then $T_k = T_c$ if $T_k < T_0$ then $T_k = T_0$
if $T_1 \geq T_2$ then $T_2 = T_k$ if $T_1 < T_2$ then $T_1 = T_k$

If there is a moisture interaction between the two points, then follow the following rules:

$$U_k = U_1 + (U_2 - 0.5 \cdot (U_1 + U_2)) \quad (10)$$

if $U_k < U_p$ then $U_k = U_p$ if $U_k > U_0$ then $U_k = U_0$
if $U_1 \geq U_2$ then $U_1 = U_k$ if $U_1 < U_2$ then $U_2 = U_k$

5. Description of the calculation algorithm

Therefore, in order to determine the values of temperature and humidity in the studied of 3D model of wood, we can use the algorithm that you can see in Figure 2.

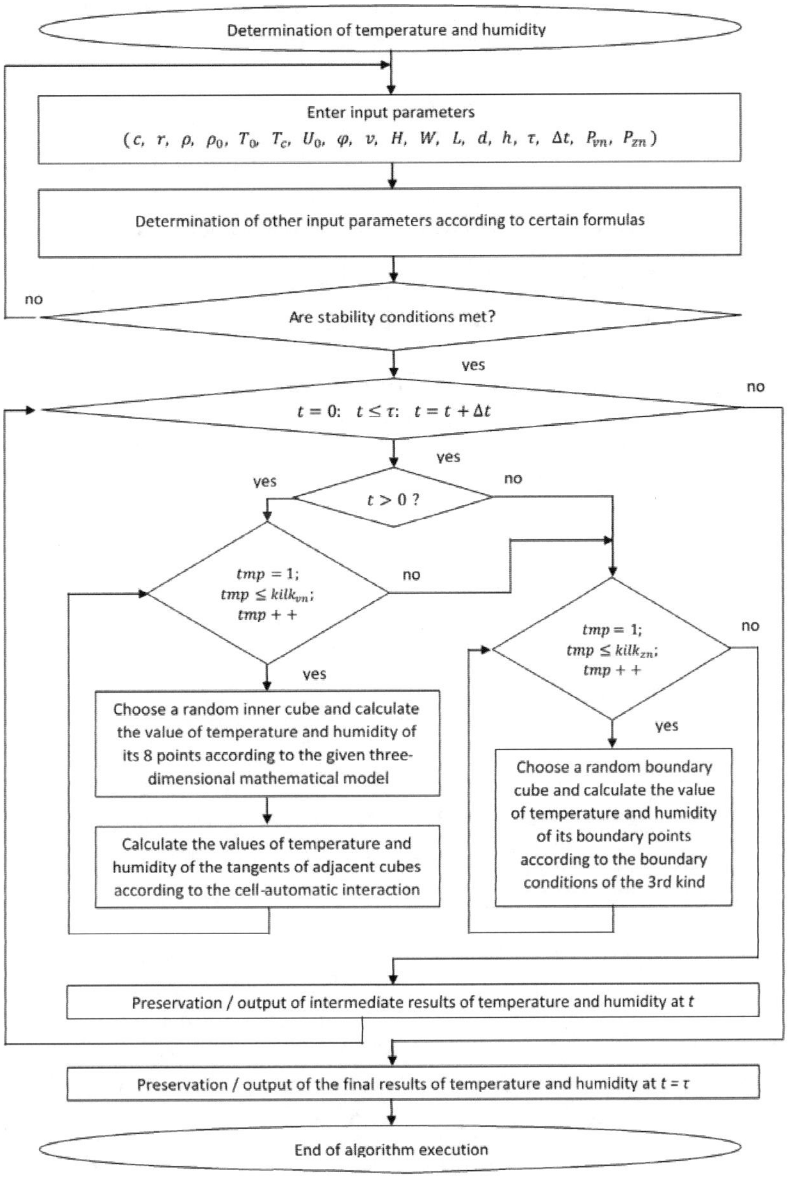

Figure 2. Algorithm for calculating temperature and humidity in 3D model of wood.

When we start calculations, we check the stability conditions for an explicit difference scheme [17].

$$\Delta t\left(\frac{\lambda_1}{h_1}+\frac{\lambda_2}{h_2}+\frac{\lambda_3}{h_3}\right)\le\frac{c\rho-\varepsilon\rho_0 r}{2}$$
$$\Delta t\left(\frac{a_1}{h_1}+\frac{a_2}{h_2}+\frac{a_3}{h_3}\right)\le\frac{(1+\delta)^{-1}}{2} \tag{11}$$

If stability conditions fulfilled, then we can start the calculation, if not, then we need to change the values of the input parameters. In calculation, process we randomly select coordinates of certain number of internal cubes, where the values of temperature and humidity at the main points will be, calculate according to a three-dimensional mathematical model (1). The values of temperature and humidity of the tangents of adjacent cubes will be determined by using the cell-automatic interaction, which described in the previous section.

In turn, the values of temperature and humidity for the points of the boundary cubes will be determined according to the boundary conditions of the third kind (4). Since we don`t use cellular automata when calculating boundary conditions, we need to use all boundary cubes to calculate.

6. Conducting a test experiment

Therefore, to verify the adequacy of the developed algorithm, we conducted a series of test experiments [18]. The main input parameters have the following values: T_0 = 30°C, T_c = 100°C, $U_0 = 0.35$, $\varphi = 0.75$, $\Delta t = 60$ sec., $\tau = 864000$ sec. In total, the studied model is represented by 9375 cubes, 6877 of which are internal cubes. By using a mathematical model, we will calculate the temperature and humidity of the points for 20% of the inner cubes. The values of temperature and humidity for the points of other inner cubes will be determined by using cellular automata.

Below you can see the results of calculations the temperature and humidity (see Figure 3) for the target point of the cube, which is in the center of the studied model. The figure on the left shows the graph of temperature distribution, and the figure on the right shows the graph of humidity distribution. The left axis is responsible for the numerical values of T or U, and the lower axis is responsible for the iteration number.

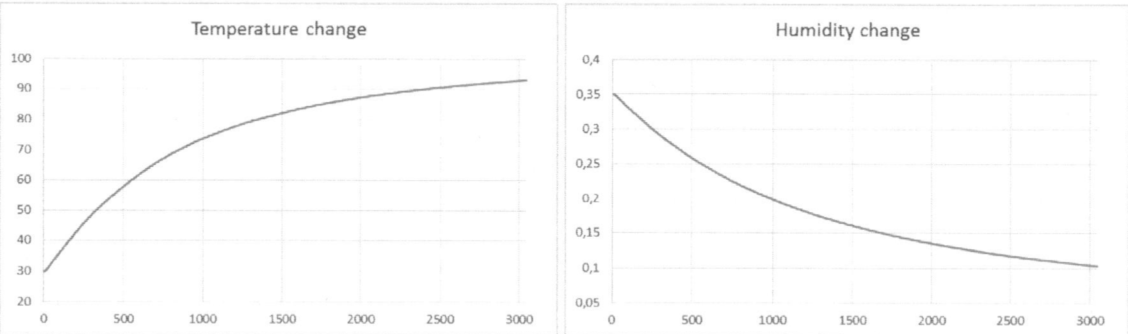

Figure 3. The results of calculations the temperature and humidity for the target point of the cube.

As a result of calculations, the target point located in the center of the studied model received the value of equilibrium humidity at 3046 iterations. That is, we need to spend 3046 * 60 sec. = 182,760 sec. (2,115 days) in order to dry the 3D model of wood of a given size. We also conducted another experiment, but without using the cellular automata. The result was found in 3019 iterations. It is worth noting that to perform the second experiment we needed about 7 minutes, while the implementation of the first experiment took us about 1.5 minutes. This proves that a cellular automata in the simulation of thermal conductivity problems of this type significantly speeds up the time of obtaining results. It is also worth noting that the use of cellular automata allows us to maintain the adequacy of calculations if more than 10% of internal cubes calculated according to a three-dimensional mathematical model (1).

Conclusions

In this work, we demonstrated the use of cellular automata in the study of heat and mass transfer processes, in particular during wood drying. During the calculations, we used a three-dimensional mathematical model, which described by a system of interconnected differential equations in partial derivatives by spatial coordinates and time. We also used cell automata during the study. To use them, we presented the studied model in the form of a set of cubes.

As a result of this work, we performed a number of experiments. Their conduction showed the possibility of using cellular automata in solving thermal conductivity problems. In addition to solving the problem of thermal conductivity, it was possible to achieve stability and accuracy of results, even with a significant increase in density and time step. In addition, the use of cellular automata allows us to significantly speed up the calculation process. This acceleration was achieved due to the ability to change the number of cubes in which the numerical values of temperature and humidity for points are determined according to a three-dimensional mathematical model. Numerical values of temperature and humidity for the points of other cubes directed to the state of equilibrium according to the total simulation time by using developed cellular-automatic interactions between the internal points of one cube and between the tangents of different cubes.

References

[1] White P V 2005 *Theoretical foundations of thermal processing and drying of wood* (Publishing House "AGE") p 364

[2] Farmaga I Lobur M Shmigelskyi P Javorskyi N and Spiewak P 2012 *Regular and adaptive meshing algorithms for modeling of spherical inclusions by finite element method* (Proceedings of International Conference on Modern Problem of Radio Engineering, Telecommunications and Computer Science, Lviv-Slavske) pp 505-507

[3] Wolfram S 2002 *A new kind of science* (Wolfram Media Inc.) p 1021

[4] Hadeler Karl-Peter and Muller Johannes 2017 *Cellular Automata: Analysis and Applications* (Springer International Publishing) p 425

[5] Jaworski N Iwaniec M and Lobur M 2019 *Implementation Features of Composite Materials Effective Mechanical Characteristics Finding Method Based on Microlevel Cellular Structural Models* 2019 IEEE 15th International Conference CADSM Polyana Ukraine pp 1-5

[6] Jaworski N Iwaniec M Andrushchak N Wozny J Lobur M and Karkulovskyy V 2020 *Simulation of Corrugated Graphene Structure by Microlevel Cellular Models and Gaussian Random Fields* IEEE 16th International Conference on the Perspective Technologies and Methods in MEMS Design (MEMSTECH) 127-132

[7] Jaworski N Andrushchak N Lobur M Iwaniec M and Sitek W 2019 *Composite material microlevel cellular model data transfer and verification by fem analysis software systems* IEEE 15th International Conference on the Perspective Technologies and Methods in MEMS Design (MEMSTECH) 17-22

[8] Jaworski N Lobur M and Iwaniec M 2018 *Implementation features of cellular composites microlevel structural models construction based on Voronoi tessellation and OpenCL technology usage* 14th International Conference on Perspective Technologies and Methods in MEMS Design (MEMSTECH) 109-113

[9] Jaworski N Iwaniec M Lobur M and Farmaga I 2016 *Diamond-based metal matrix composite effective heat conduction coefficient synthesis microlevel model verification* 12th International Conference on Perspective Technologies and Methods in MEMS Design (MEMSTECH) 25-27

[10] Toffoli T and Margolus N 1991 *A Cellular Automata Machine* (World) p 280

[11] Sokolovskyy Ya and Sinkevych O 2018 *Software and algorithmic support for represettation of CAD models in 2D von Neumann neighborhood* **2300** (CEUR Workshop Proceedings) pp 215-218

[12] Kireeva A 2013 *Parallel implementation of totalistic cellular automata model of stable patterns formation* **7979** (Parallel Computing Technologies) pp 347-360

[13] Sokolovskyy Ya Sinkevych O and Voliansky R 2019 *Development the software for simulation of physical fields in wood drying chambers by using cellular automata* (15th International Conference CADSM 2019 Polyana Ukraine) pp 24-27

[14] Sokolovskyy Y and Sinkevych O 2017 *Calculation of Drying Agent in Drying Chambers* (XIV International Scientific and Technical Conference CADSM 2017 Polyana Ukraine) pp 27-31

[15] Mazumder Sandip 2016 *Numerical Methods for Partial Differential Equations, Finite Difference and Finite Volume Methods* (Academic Press Elsevier Inc. Langford Lane UK) p 464

[16] Shumylyak L Zhikharevich V and Ostapov S 2016 *Modeling of impurities segregation phenomenon in the melt crystallization process by the continuous cellular automata technique* **290** (Applied Mathematics and Computation) pp 336-354

[17] Nikitenko N I 1983 *Theory of heat and mass transfer* (Science opinion) p 352

[18] Sokolovskyy Y Sinkevych O and Voliansky R 2019 *Software for Studying Wood Drying Chambers Based on SolidWorks Flow Simulation Experiment* (International Conference ACIT 2019 Cesce Budejovice Czech Republic) pp 281-284

IOP Conf. Series: Materials Science and Engineering 1016 (2021) 012018 doi:10.1088/1757-899X/1016/1/012018

Issues with Belt Guidance and Device Controls in the Automated Conveyor and Drive Belt Perforation Process

D Wojtkowiak[1], K Talaśka[1], G Domek[2*], D Wilczyński[1] and J Górecki[1]

[1] Institute of Machine Design, Poznan University of Technology, Piotrowo 3 street, 61-138 Poznań, Poland
[2] Kazimierz Wielki University in Bydgoszcz, Faculty of Mathematics, Physics & Technical Sciences, 85-092 Bydgoszcz, Poland

*Email: gdomek@ukw.edu.pl

Abstract. The perforation process for multi-layer composite polymer belts presents numerous challenges due to the varied mechanical characteristics and geometrical parameters of the entire range of belts used in the industry. Many problems can be minimized or eliminated with correct development of construction features of the machines for high precision mechanical belt perforation. Regardless of the construction solutions employed in the devices, a major factor influencing the quality of the final product as well as machine reliability is the method of controlling their operation as well as the correct guidance of the belt throughout the entire flow of material in the manufacturing process. The present work provides an analysis of the assemblies ensuring correct belt guidance and presents the causes and effects of incorrect development of these systems. Furthermore, it touches upon the aspects involving the control solutions for such equipment involving the definition of actuators and controllers, the selection of sensors and methods of measuring linear displacement of the belt and perforating head position together with the control algorithms for carrying out different patterns of perforation.

1. Belt guidance in the perforation process

The perforation process for multi-layer composite polymer belts presents numerous challenges due to the varied mechanical characteristics and geometrical parameters of the entire range of belts used in the industry [1, 2]. Many problems can be minimized or eliminated with correct development of construction features of the machines for high precision mechanical belt perforation. Regardless of the construction solutions employed in the devices, a major factor influencing the quality of the final product as well as machine reliability is the method of controlling their operation as well as the correct guidance of the belt throughout the entire flow of material in the manufacturing process.

Belt guidance in the course of the perforation process should be carried out along the entire length of the manufacturing line, from the belt unreeling system and any buffer that may be employed, to prevent continuous operation, through the feeding system which needs to accurately introduce the material to the perforating system, to the module which collects the finished perforated belt. It is critical to maintain the base side alignment at the entire length of the belt, as even a small deviation of approx. 1 mm at the distance of several dozen meters causes to exceed the allowable standard tolerances and destroys the belt. To this end, various gates and stops are to be employed to prevent excessive transverse motion of the belt. At the same time, it should be considered that the belt width may vary by at least a few millimetres along its length; therefore, one needs to facilitate automatic adjustment so that the belt is not blocked or wrapped.

The critical supports are the belt edge grippers (Fig. 1) present in the perforation module MP, directly in front and after the perforating head. Apart from maintaining the base side alignment of the belt, they are to prevent its motion following a transverse movement of the perforating heads. The base of the gripper uses profile 1 with the defined height, with the bottom plate 2 being mounted at its face surface and the plate acting as a sliding surface for the belt. Bottom plate 2 is mounted to the angle piece 3 with bolts 4, which is subsequently fastened to profile 1 using bolt 5 and t-slot nut 6 placed in the groove R of profile 1. The pneumatic cartridge actuator 7 is fastened to the upper part of the angle piece, secured with two nuts 8 on both sides of the angle piece, which also facilitate its vertical adjustment. At the end of the piston rod 7a of the cartridge actuator 7 a disc gripper 9 is fastened and secured with a lock nut 10. The side surface of the bottom plate 2 allows to fasten the optical gate BO using the mounting plate 11, which allows to adjust the extension of the gate which affects the precision of the perforation pattern, and bolts 13. The profile 1 is fastened to the frame of the perforation module MP using a mounting angle 12 facilitating the adjustment of the gripper in relation to the MP frame. The extension of the actuator is affected by the electro valve. The operating cycle of such gripper entails pressing the belt down as the set longitudinal position is achieved and maintaining this position until the belt moves so that another row of the perforation pattern can be executed.

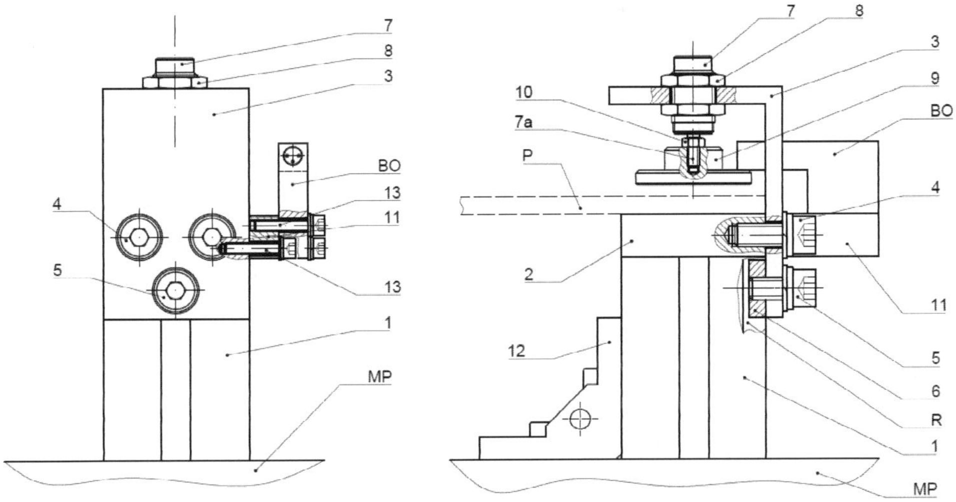

Figure 1. Belt edge gripper construction in an automatic device for high precision belt perforation: 1 – base, 2 – bottom plate, 3 – angle piece, 6 – t-slot nut, 7 – cartridge actuator, 7a – actuator piston rod, 8 – nut, 9 – pressing disc, 10 – lock nut , 12 – connecting angle, 4, 5, 11, 13 – mounting bolts, BO – optical gate, MP – perforation module

Maintaining insufficient base side alignment results in deviations from the programmed perforation pattern and may cause the belt to twist or block, damaging the product. The optical gate BO is used to detect the presence of the belt, and in the case of misalignment of the base side exceeding the positioning deviation value (e.g. 0.5 mm), the process must be stopped so that the belt isn't damaged. Additionally, the gate is used to detect the beginning of the belt which improves the belt positioning repeatability. If the process is stopped as a result of malfunction or incorrect belt guidance, it is possible to resume it after the belt is remounted on the machine.

Another major issue is the slip between the driving rollers of the feeding system and the transported belt. This slip is different for different belt types, flat or toothed, and affects feeding accuracy. Therefore, it is necessary to adjust the pressure applied to the belt, and the system should facilitate uniform distribution of the contact stress in order to prevent belt chamfering as a result of velocity vector occurring in the transverse direction in relation to the belt's motion. Too low pressure on the belt results in no friction coupling with the driving roller; consequently, the motion of the belt is not carried out. Excessive pressure on the belt causes deformation, which can result in a jerking motion (particularly with toothed belts). Based on the studies carried out during pilot production, the optimal pressure values were determined as 1.5 bar for polyurethane toothed belts and 5-6 for coated flat belts, with external layer of polyurethane, nitrile rubber, plastic fabric or PVC, with actuators with piston diameter equal to 50 mm.

To prevent material waste due to incorrect belt guidance, numerous control systems are utilized to monitor the presence of the belt side and measuring its linear displacement at several points along the length of the manufacturing line. Such a solution enables to compare the values and correct the alignment as necessary so that the programmed pattern of perforation is achieved. Compensating for positioning error entails calculating the difference between the set displacement and the displacement read by the encoder achieving friction coupling with the belt. The difference in value is sent to the BLDC drive controller operating in closed loop in incremental mode. The motion is carried out until the controller receives information about the drive finishing the motion, at which point the difference is determined once again. If the difference does not exceed the threshold value (0.5 mm), the belt side grippers close and the row of holes is being punched. The construction of the device for measuring linear displacement of the belt is shown on Fig. 2.

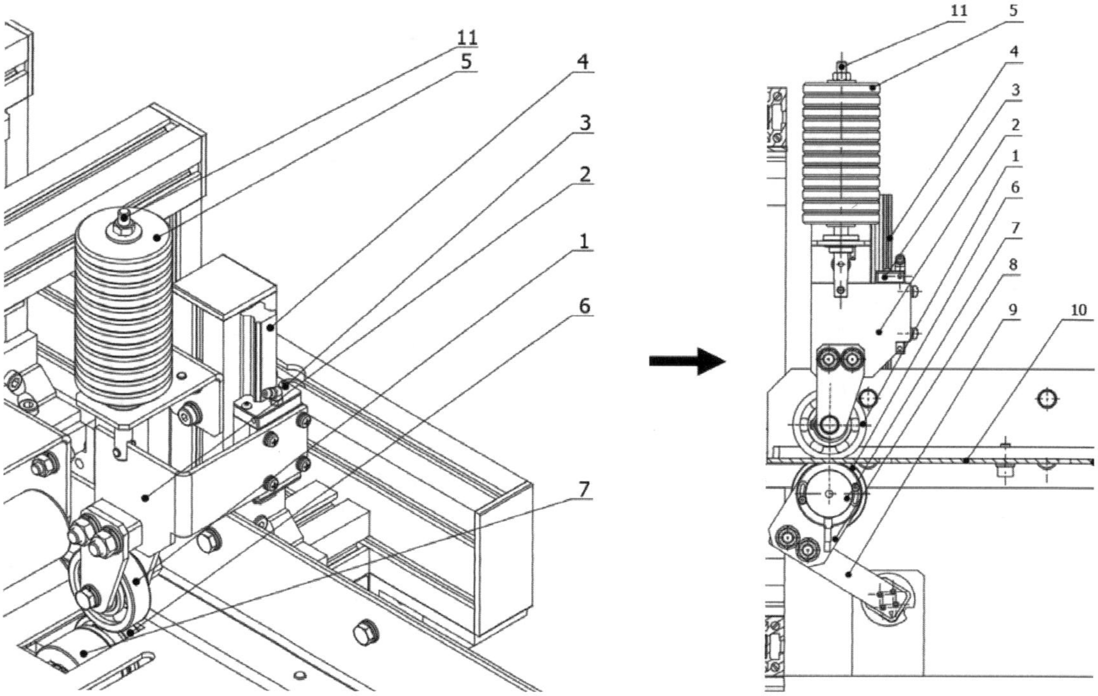

Figure 2. Construction design of the device for measuring linear displacement of the belt when carrying out perforation: 1–pressure roller, 2 – roller mounting, 3 – carriage, 4 – guide rail, 5 – weight, 6 – measuring roller, 7 – rotation angle sensor, 8 – sensor mounting, 9 – arm, 10 – plate, 11 – axle

The device for measuring belt displacement consists of two rollers: upper roller with bearing and lower rotation sensor roller. The cylinder surfaces of both rollers are in contact with, respectively, the top and bottom surface of the belt. The upper roller with bearing is attached to the device body which is connected to the carriage which moves along the vertical guide to facilitate adjustment of its height, enabling to introduce belts of any thickness between both rollers. Moreover, the upper body to which the upper roller with bearing is connected is fastened to an axle, on which weights can be placed to facilitate variable adjustment of the pressure of the upper roller with bearing onto the top surface of the belt, consequently ensuring the belt is indirectly pressed to the lower rotation sensor roller. This allows eliminating the slipping of the belt surface during linear motion in relation to the rotation sensor roller surface. The displacement of the belt causes both rollers to rotate, whereas the rotation of the bottom roller is registered by the rotation sensor which allows determining the linear displacement of the belt. The bottom roller is set on the axle of the rotation sensor which is connected to an arm which facilitates angular motion around an axis at one of the ends. Forced angular deflection of the arm is a reaction associated with the arm attempting to assume the starting angular position. This allows eliminating the risk of applying too much pressure on the belt and the rotation sensor roller, which might damage the sensor. The rotating arm discussed above is connected to the base and rotates along an angle. The base is connected to the body of the device in which the belt displacement is to be measured. The measured belt moves on a plate while perfectly adjacent to it, therefore the lower roller of the rotation sensor is extended through a special opening in the plate, slightly above the surface which the belt adheres to.

The design aspects presented above may significantly affect the automated perforation process. However, one needs to point out that they are used in conjunction with control algorithms, which may also cause difficulty in achieving correct pattern of perforation.

2. Controlling the automatic belt perforation machine

Among the subjects related to automatic belt perforation machine controls, one can highlight the following issues: the PI or PID controller settings for used drive systems which enable to achieve the desired process dynamics while maintaining positioning accuracy, perforating head motion sequence programming which minimizes the number of motions and significantly affects the perforation process as well as the operator-machine interface via HMI panel, employment of the required sensors, communication protocols, data acquisition methods and programming the specific perforation patterns.

Fig. 3a demonstrates an example implementation of an automatic belt perforation machine control system. The main component is the programmable controller with HMI operator panel. The drive solutions for perforating heads, the operating drive unit (ZNR) and drive rollers for belt feeding utilize BLDC drive motors together with regulators and controllers communicating via ETHERNET/IP protocol. Belt unreeling and winding is achieved through vector inverter controlled motoreducers, with communications achieved via RS485, additionally, the modules can be operated manually with pedal switches. Controlled pneumatic pressure systems comprise of pneumatic actuators controlled by 3/2 bistable electro valves, whereas Hall effect sensors are installed on actuators in order to detect extreme piston positions. The working units for perforation and cutting with pneumatic actuators are controlled by 5/3 electro valves with two solenoids, with extreme positions being signalled by Hall effect sensors installed on actuator enclosure. Reference positioning of heads and ZNR uses 6 induction sensors acting as limit switches. Additionally, for the purpose of monitoring the state of buffers and module controls for reeling in and unwinding utilizes an additional 8 K1-K8 sensors placed according to the diagram provided on Fig. 3b.

To ensure process continuity without the need to synchronize individual drives, the following states and operations were assigned to individual sensors:

- K1 – empty input buffer → stop perforation and refill buffer,
- K2 – minimum input buffer fill during perforation → restart perforation / begin to fill the buffer,
- K3 – input buffer full → stop operation of unreeling modules,
- K4 – no material in input buffer,
- K5 – output buffer empty → stop operation of the winding buffer,
- K6 – maximum input buffer fill during perforation → restart perforation / begin to unload buffer,
- K7 – output buffer full → stop perforation and empty the buffer,
- K8 – no material in output buffer.

A critical factor influencing the work quality of the perforation machine is the positioning accuracy of the belt and perforating heads. Therefore, it is necessary to determine the controller settings through experimental studies assisted by the drive system dynamics model [3]. In the developed control system, PI (proportional-integral) regulators [4] were used to control both the displacement and speed. After a series of trials, optimal parameters were selected for the designed construction. However, it should be considered that depending on the set displacement, the displacement error and control

IOP Conf. Series: Materials Science and Engineering 1016 (2021) 012018 doi:10.1088/1757-899X/1016/1/012018

overshoot will be different. Based on the results presented on Fig. 4, one can determine that for the displacement value of x = 10 mm, the adjustment error stabilizes at approx. e = 0,75 mm and the overshoot p = 2 mm, for x = 50 mm, e = 1 mm p = 3,3 mm, and for x = 100 mm, e = 1,8 mm p = 5 mm. Consequently, depending on the programmed perforation pattern, it is advantageous to adjust the PI regulator parameters e.g. by employing the characteristics determined on the basis of machine tests. During testing, it was observed that increasing the value of the proportional term P affects the positioning error and overshoot value, whereas the setting of the integral terms I affects the stabilization time of the positioning error. Extreme regulator setting values are to be avoided as too high values introduce a resonance in the system which may damage the device.

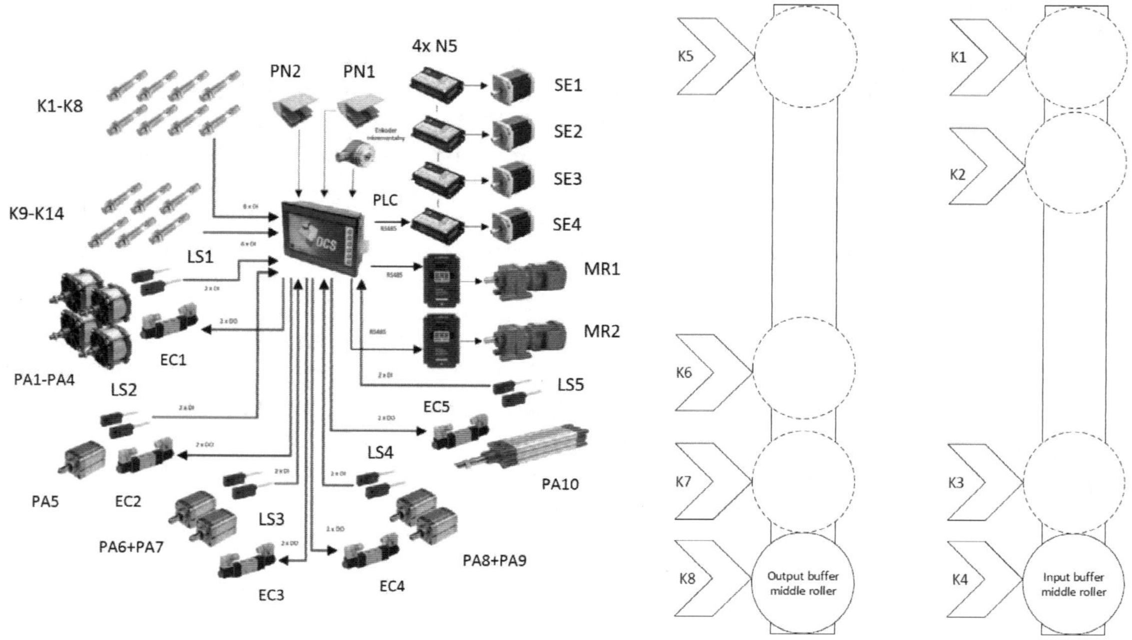

Figure 3. Example implementation of the control system for automatic belt perforation: a) conceptual diagram, b) placement of sensors in buffering modules: K1-K14 – inductive proximity sensors, LS1-LS5 – magnetic piston positioning sensors, EC1-EC5 – electrovalves with solenoids, PA1-PA10 – pneumatic actuators, MR1-2 – motoreducers, SE1-4 – BLDC electric motors, PLC – programmable logic controller, 4×N5 - BLDC electric motors regulators, PN1-2 – pedal switches

A critical factor influencing the work quality of the perforation machine is the positioning accuracy of the belt and perforating heads. Therefore, it is necessary to determine the controller settings through experimental studies assisted by the drive system dynamics model [3]. In the developed control system, PI (proportional-integral) regulators [4] were used to control both the displacement and speed. After a series of trials, optimal parameters were selected for the designed construction. However, it should be considered that depending on the set displacement, the displacement error and control overshoot will be different. Based on the results presented on Fig. 4, one can determine that for the displacement value of x = 10 mm, the adjustment error stabilizes at approx. e = 0,75 mm and the overshoot p = 2 mm, for x = 50 mm, e = 1 mm p = 3,3 mm, and for x = 100 mm, e = 1,8 mm p = 5 mm. Consequently, depending on the programmed perforation pattern, it is advantageous to adjust the PI regulator parameters e.g. by employing the characteristics determined on the basis of machine tests. During testing, it was observed that increasing the value of the proportional term P affects the positioning error and overshoot value, whereas the setting of the integral terms I affects the stabilization time of the positioning error. Extreme regulator setting values are to be avoided as too high values introduce a resonance in the system which may damage the device.

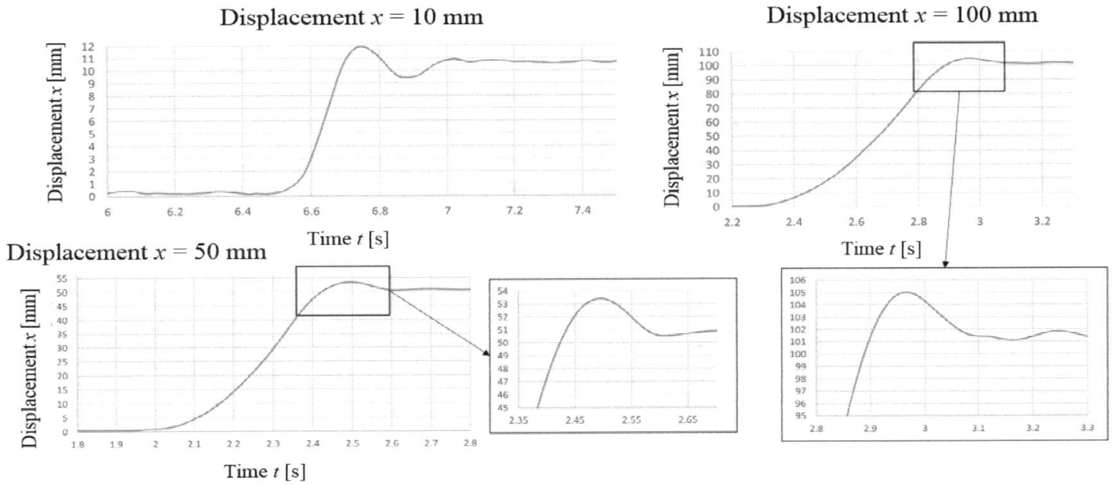

Figure 4. Measurement results of the perforation head displacement – response from the regulation system to the set input function

To facilitate carrying out any perforation pattern programmed by the user, it is necessary to develop complex algorithms for calculating the coordinates for subsequent holes, while keeping in mind the requirement to minimize the motions of the head. The present work discusses machine controls used to achieve one of the two basic patterns of perforation: rectangular (A) or displaced (B), presented on Fig. 5 together with programmable parameters.

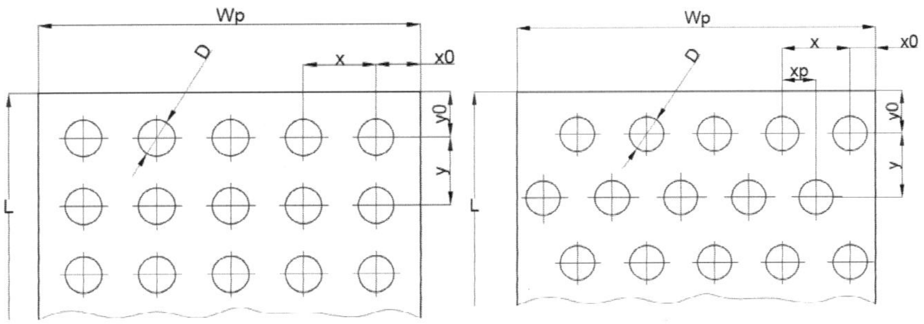

Figure 5. Rectangular and displaced patters of perforation: *Wp* – belt width, *L* – belt length, *D* – hole diameter, x - distance between neighboring hole columns, *x0* – distance between the first column from belt side, *xp* – displacement value between columns in neighbouring rows *y* – distance between neighboring rows of holes, *y0* – distance of the first hole in the row from belt side [5]

The presented control method (Fig. 6 and 7) is based on calculating hole coordinates and operating in an absolute system of reference in which the zero point is determined at every startup of the machine using limit sensors, through free motion of the heads until the sensor is triggered. One needs to consider that the travel speed is highly important from the standpoint of transverse positioning. For each row, the controller calculates the coordinates and sends them in a sequence to the regulator unit of the respective drive motor, until the *i* incremental counter reaches the value equal to the number of columns in the perforation pattern. In the case of pattern A, the coordinates do not change, whereas for pattern B, every second row has different coordinates. Displaced patters are also used, in which every other row has one fewer hole. To achieve such a pattern, the controller checks the reminder of the division of the current row number by 2, and subsequently, depending on the result, instructions are carried out from branches A or B, where the displacement value x_p is taken into account, or condition to end the conditional instruction from „equal to or lower than" to „lower than". Additionally, the controller constantly monitors the buffering system, operating the systems for unwinding and reeling in at branches N and O. In order to make the hole, the electrovalve is engaged to switch the solenoids after the signal from actuator end sensor is received.

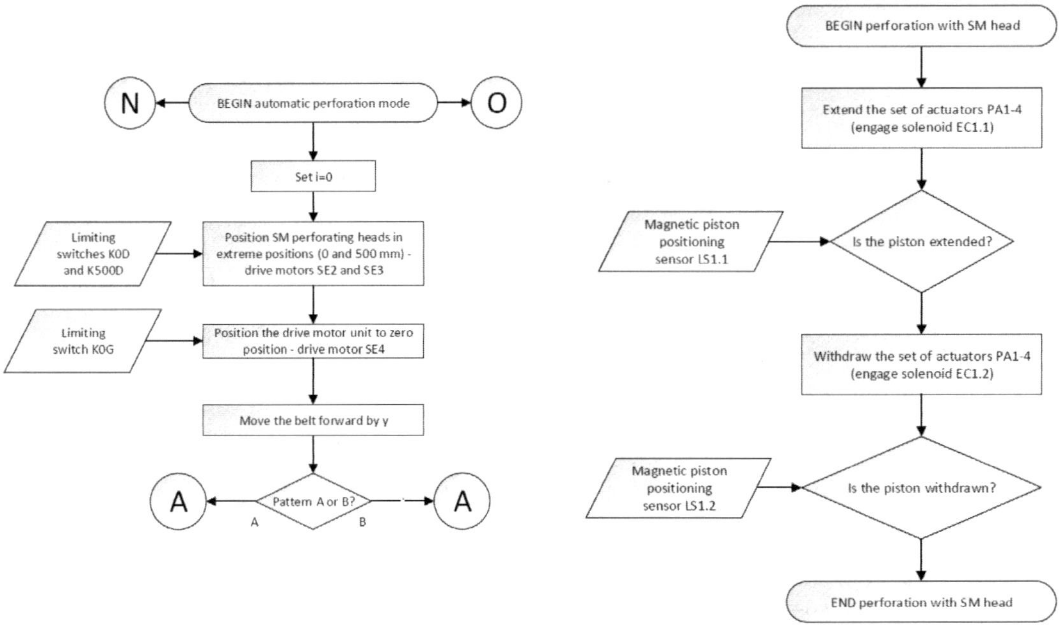

Figure 6. Control algorithm for belt perforation device according to patterns A and B – part 1

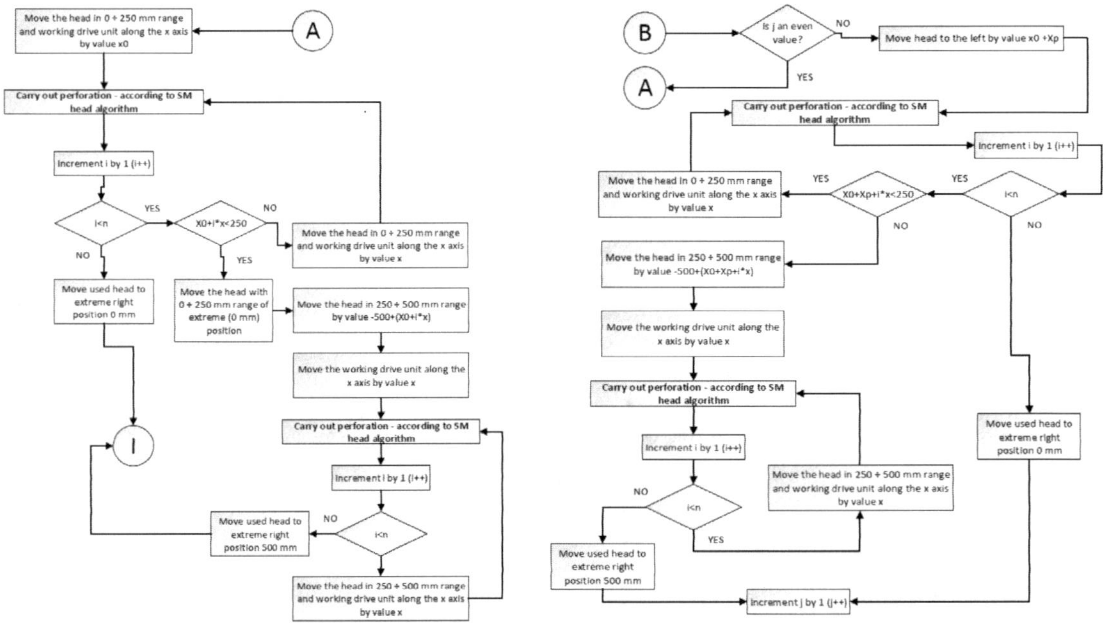

Figure 7. Control algorithm for belt perforation device according to patterns A and B – part 2

Control algorithms for more complex patterns, e.g. double row or interrupted patterns, can be developed in a similar way. If the peroration pattern calls for punching holes with different diameters, a repeatable pattern of positioning is necessary, so all the holes with a given diameter can be made in a single pass. All the above mentioned aspects necessitate a long startup time for the machine, despite the deceptive simplicity of the process, as multiple tests are required to verify the degree of achieved accuracy.

Conclusions

The issues described in the present work confirm that within a mechatronic system, the discussed design being such, both the construction design and device controls are important for the final result of the machine operation. It needs to be pointed out that both issues are difficult to separate as they will affect one another to a great extent. However, by following the recommendations provided in this work, it is possible to significantly improve the design process for the control systems of these machines and the presented issues can be applied to other machining processes involving belts and strips.

References

[1] Wojtkowiak D and Talaśka K 2019 Determination of the effective geometrical features of the piercing punch for polymer composite belts, *Int. J. of Adv. Manuf. Technol.* **104** 315–32

[2] Wojtkowiak D, Talaśka K, Malujda I and Domek G 2018 Estimation of the perforation force for polymer composite conveyor belts taking into consideration the shape of the piercing punch, *The Int. J. of Adv. Manuf. Technol.* **98**:9-12, 2539–61

[3] Andreev A and Peregudova O 2018 Non-linear PI regulators in control problems for holonomic mechanical systems, *Syst. Sci. & Contr. Eng.* **6**:1, 12-9

[4] Rumniak P, Ufnalski B and Grzesiak L 2015 Tuning of PI regulators in distributed control system for an electric vehicle, *Przegląd Elektrotechniczny* **91**:**9**, 290-94

[5] Wojtkowiak D and Talaśka K 2018 Problematyka procesu perforacji pasów transportowych za pomocą stempla i matrycy, *Inżynieria Wytwarzania* (Wyd. PWSZ w Kaliszu) pp 251-64

Use of Methods and Technologies of Additive Production for Optimization of Parameters of Designs

A Zdobytskyi[1]*, M Lobur[1], T Klymkovych[1], R Kaczynski[2] and A Vasiliev[3]

[1] CAD Systems Department, Lviv Polytechnic National University, Mytropolyta Andreya St., Lviv, 3 Ukraine
[2] Bialystok University of Technology, Wiejska St., 45a, 15-351 Białystok, Poland
[3] Chair of Theory and Computer-Aided Design of Mechanisms and Machines, National Technical University, Kharkiv Polytechnic Institute, Kyrpychova St., 2, Kharkiv, Ukraine

*Email: andrii.y.zdobytskyi@lpnu.ua

Abstract. The article presents the existing methods of implementation of rapid prototyping technologies in rapid production. It is performed using computer modeling programs with integrated modules of generative design based on different topology optimization methods. In the environment of ANSYS Mechanical (Workbench) the finite element method established the dependence of strength parameters on the density of the generated mesh of the reinforcing contour of the part. According to the results of the simulation, the graphical dependencies of the maximum general stresses and the margin of safety of the part depending on the materials of the 3D models under the action of equivalent stresses were obtained.

1. Introduction.

The rapid development of scientific and technological progress requires any large and small series production to be constantly and timely updated and keep up with the times, as today's industry requires the use of newer and more advanced production technologies focused on the widespread implementation of robotic systems to control all production processes.

In turn, the gradual improvement and modernization of production require the use of more modern design methods and materials [1-3] to increase the durability and ergonomics of products. As a result, it requires the use of state-of-the-art production methods, one of which is the rapid prototyping technology of Industry 4.0 [4], which, unlike traditional production technologies, allows us to design and manufacture a production facility in the shortest possible time. Also with the help of modern methods of computer modeling, it is possible to identify weaknesses and imperfections of the structure directly at the stage of their design in terms of strength and durability (additive production) figure 1.

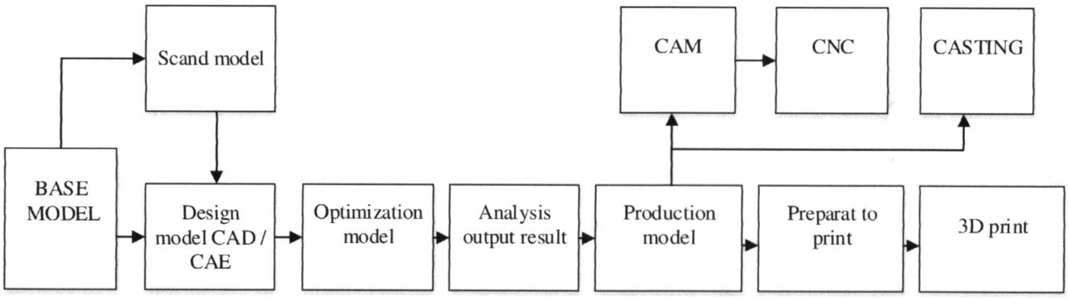

Figure 1. Scheme of the production process of detail development

Creating a 3D model or its reengineering is done in several common ways, including the use of a 3D scanner that analyzes the physical object and based on the obtained data creates a 3D model of the design object mostly in formats (STL, OBJ, PLY, WRL). The advantages of the 3D scanner are that in a relatively short time you can get an exact copy of the object (model), which is further refined in computer 3D graphics programs (Autodesk Inventor, DS SolidWorks, DS CATIA, Ascon Kompas-3D, Siemens NX, and others).

Also, to create a new, more original or improved model, computer modeling programs are used, which contain integrated modules of engineering calculation, analysis, and simulation of physical processes (FEA module) performance of the design object (dynamic simulation). As well as possible optimization of parameters of detail by mass and shape using modules shape generator, topology study, and generative design figure 2 without significantly reducing its strength characteristics [5, 6].

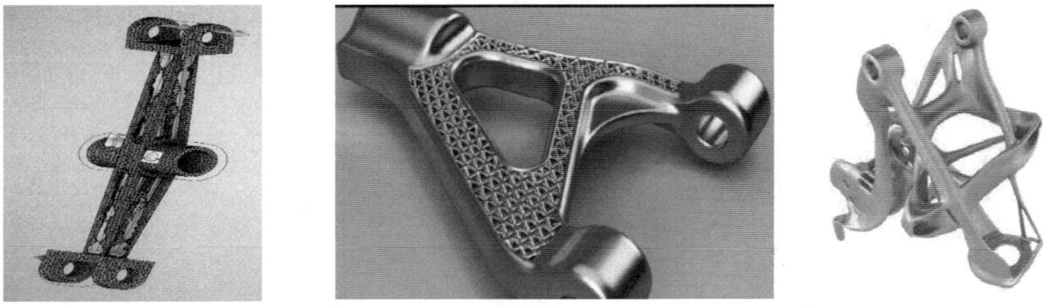

a b c

Figure 2. The ways of optimization of models by shape and mass:
a – density-based only, b –lattice-based, c – density based + generative design

To optimize the parameters of the structure and reduce its material consumption quite often there is a need to create a hollow object of complex shape, the manufacture of which on CNC machines is quite a difficult task, and sometimes impossible.

Quite often casting is used as the means of manufacturing of parts, which allows getting parts of good quality using a small amount of material. However, this method is more efficient for mass production of parts of simple shape. There are also increased requirements for the technology of manufacturing parts, as due to casting under pressure of composite materials there are various defects [7].

Also, one of the promising methods of making an object based on previously created triangulated models is its printing (growing) on a 3D printer. In turn, the technology of the production of ultra-complex elements of parts is most likely to be made by one of the technologies of 3D printing.

2. Statement of the problem

In the era of development and implementation of industry 4.0 technologies, one of the priority areas of which is 3D printing, there is a gradual requirement for the use of heterogeneous materials to create authentic and original parts by purpose and form. In particular, the technology of laser 3D printing of powder and photopolymer materials (SLM, SLS, SLA, DMT, DMD) is one of the most promising but expensive technologies. Nevertheless, 3D printing technologies of various composite materials, such as FDM, MJM, PolyJet, CJP, have become widespread. As a result, the implementation of additive production technologies allows the development of this area due to the inclusion in the plastic materials' different physical and mechanical properties, resulting in an increased melting point and, accordingly, fatigue strength characteristics of the products [4].

However, the use of various types of reinforcing materials in the plastic structure leads to rapid wearout or violation of the geometry of the printing brass nozzle, which in turn affects the accuracy (precision) of the obtained parts. The use of a nozzle made of a harder material reduces its thermal conductivity, which negatively affects the flow of the filament.

Therefore, to increase the fatigue strength characteristics of parts made of various composite materials, it is advisable to strengthen them. However, the creation of a solid frame of the part helps to increase the strength of the whole part but does not take into account the direct strengthening of the structure of the part in places of greatest stress concentration [8, 9].

Since, today in the technologies of three-dimensional CAD modeling and engineering calculation CAE systems have widely developed technologies of generated design and generative design, the use of which, at the stage of preparation of the model for 3D printing, makes it possible to generate a framework for the whole model or only its necessary part. The combination of generative design and design features of parts will help optimize their size-mass and fatigue-strength characteristics.

3. Methodology of research

To verify the adequacy of the described assumptions for parametric optimization, we use the ANSYS Mechanical (Workbench) software environment for engineering modeling and calculation of 3D models by the finite element method (figure 3, 4) [10-15].

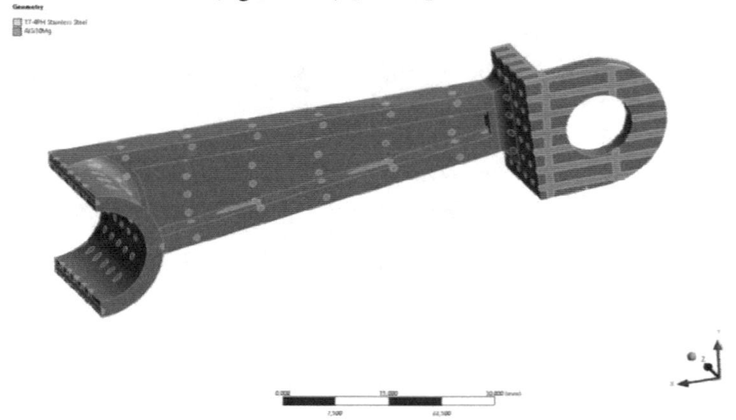

Figure 3. General view of the designed part model in ANSYS Mechanical (Workbench) environment

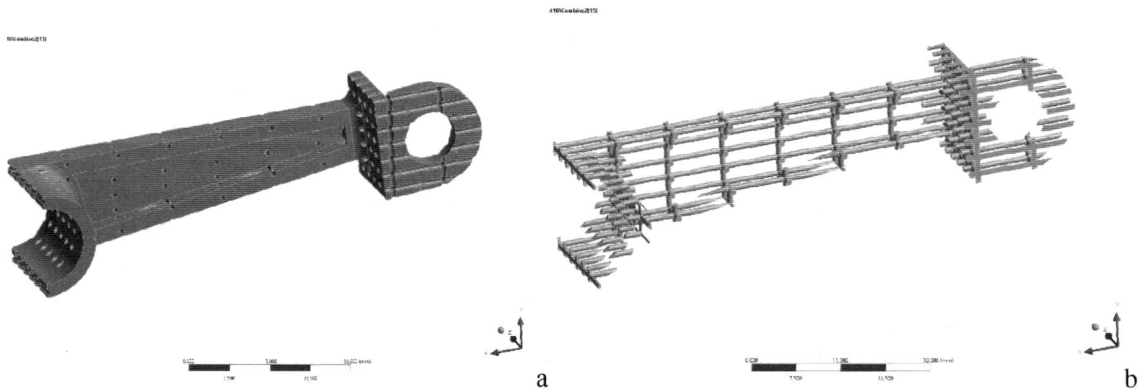

a b

Figure 4. Elements of the created design of the part model:
a - designed 3D model; b - the generated frame of the 3D model.

Characteristics of the materials which are most frequently used for production details of 3D printing (see table 1).

Table 1. The main characteristics of the used material

Material		Elasticity (Pa)	Poisson Ratio	Density (kg/m^3)	Tensile Yield Strength (MPa)		
					Current	Minimum avg.	Maximum avg.
ABS plastic	P	2,4E9	0,4	1050	36	30	50
AlSi10Mg	A	7,1E10	0,33	2670	280	100	320
Steel 316L	S	2E11	0,3	7550	230	185	800

In order to research the connection between detail materials, geometrical parameters of the reinforcement frame figure 5, and strength indicators of a 3D model plan-matrix of experimental research was compiled (see table 2).

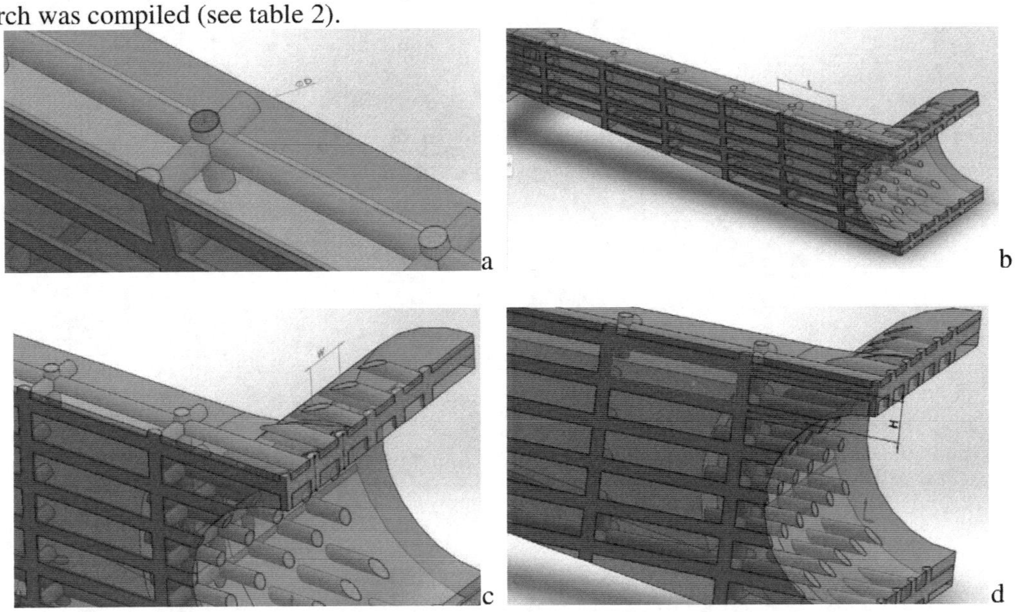

Figure 5. Dimensional parameters of the reinforcement frame:
a – diameter of the grid (d),
b – the distance between the horizontal elements of the grid in the longitudinal direction (L),
c – the distance between the horizontal elements of the grid in the transverse direction (W),
d – the distance between the vertical elements of the grid (H).

Table 2. Matrix of experimental research

	Materials*		Geometric (mm)			
	Base	Reinforcement	D	L	W	H
B	P	P	1	10	3	3
C	A	A	1	10	3	3
D	S	S	1	10	3	3
E	P	A	1	10	3	3
F	P	S	1	10	3	3
G	P	A	0.5	10	3	3
H	P	A	1.5	10	3	3
I	P	A	1	5	3	3
J	P	A	1	15	3	3

To study the details used a physical process that describes the equation (1) [10, 12]:

$$0 = \nabla \cdot s + F_V , \quad (1)$$

where F_V - the volume force vector [N/m^3], S - the Piola-Kirchhoff stress tensor

The material of the model is considered linear elastic material, which is described by the equations

$$\begin{aligned} &0 = \nabla \cdot s + F_V \\ &S = S_{ad} + C : \epsilon_{el}, \ \epsilon_{el} = \epsilon - \epsilon_{inel} \\ &\epsilon_{inel} = \epsilon_0 + \epsilon_{ext} + \epsilon_{th} + \epsilon_{hs} + \epsilon_{pl} + \epsilon_{cr} + \epsilon_{vp} \\ &S_{ad} = S_0 + S_{ext} + S_q \\ &\epsilon = \frac{1}{2}\left[(\nabla u)^T + \nabla u\right] \\ &C = C(E, \nu) \end{aligned} \quad (2)$$

IOP Conf. Series: Materials Science and Engineering 1016 (2021) 012019 doi:10.1088/1757-899X/1016/1/012019

We believe that the part is fixed on the inner side of the central hole (figure 6), i.e. on the specified surface the condition is met $u = \underset{\cdots}{0}$, where u is a moving vector.

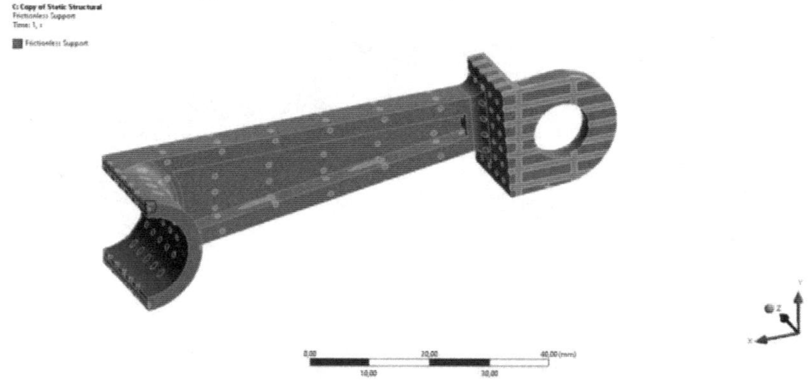

Figure 6. Rigidly fixed surface.

Also, the body is subjected to the gravitational force, as well as the force applied to the inner surfaces of the side holes (figure 7), which is described by equation

$$S \cdot n = \underset{\cdot\cdot}{F}_A$$

$$F_A = \frac{\underset{\cdot\cdot}{F}_{tot}}{A} \tag{3}$$

where F_A - force per unit area [N/m^2], F_{tot} - total force [N], A – area [m^2]

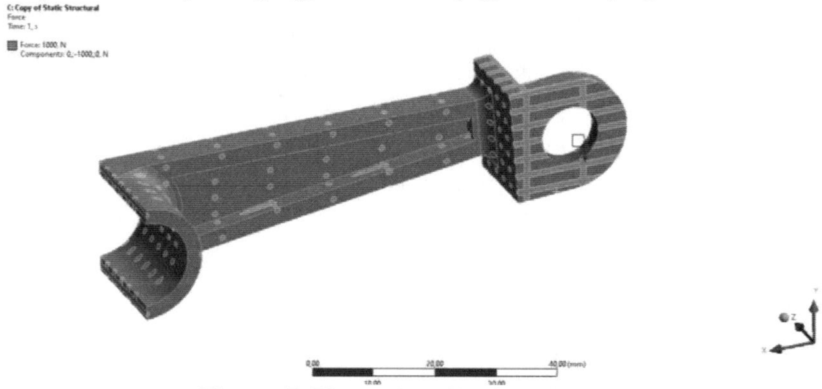

Figure 7. Place of application of force.

Total force applied along the y component axis and is 1000 [N]. Mesh – Average Element size – from 0,5 to 1,5 mm, element type – solid, parabolic. The main parameters of the division into finite elements are given in the figure. 8

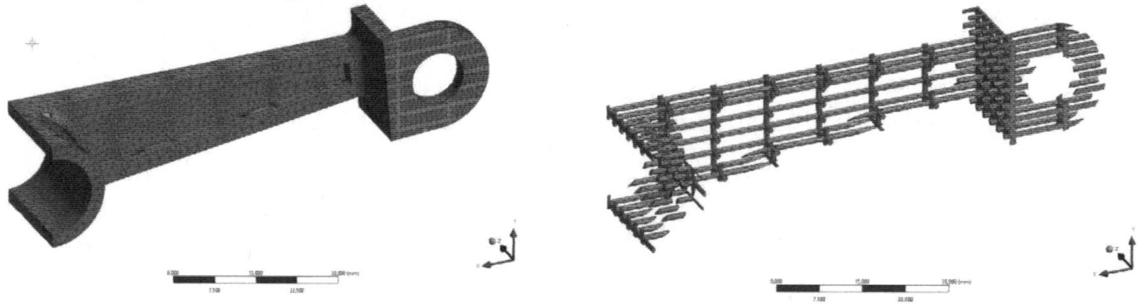

a b

Figure 8. Partitioning into finite elements:
a – designed 3D model
b – generated frame of a 3D model

IOP Conf. Series: Materials Science and Engineering 1016 (2021) 012019 doi:10.1088/1757-899X/1016/1/012019

4. Simulation results

Based on the stationary and frequency research, the following simulation results are obtained, which are presented in figures 9-13.

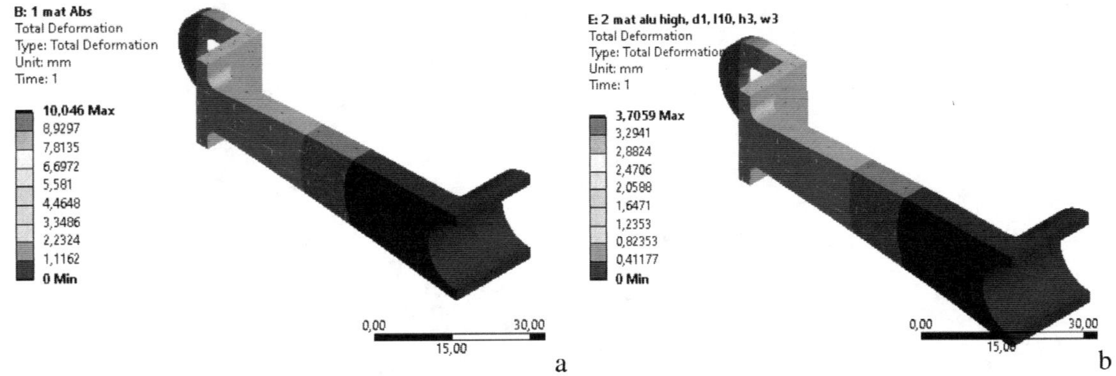

a b

Figure 9. The results of studies of total deformation
a - ABS, without reinforcement,
b - ABS Base, with aluminum reinforcement

As can be seen from figure 9, total deformation is three times smaller using a reinforcement frame

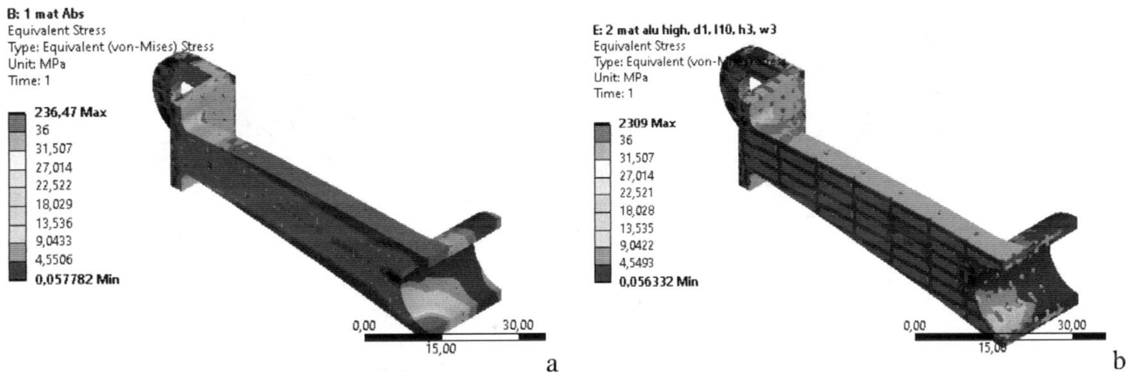

a b

Figure 10. Stress (Full construction):
a - ABS, without reinforcement,
b - ABS Base, with aluminum reinforcement

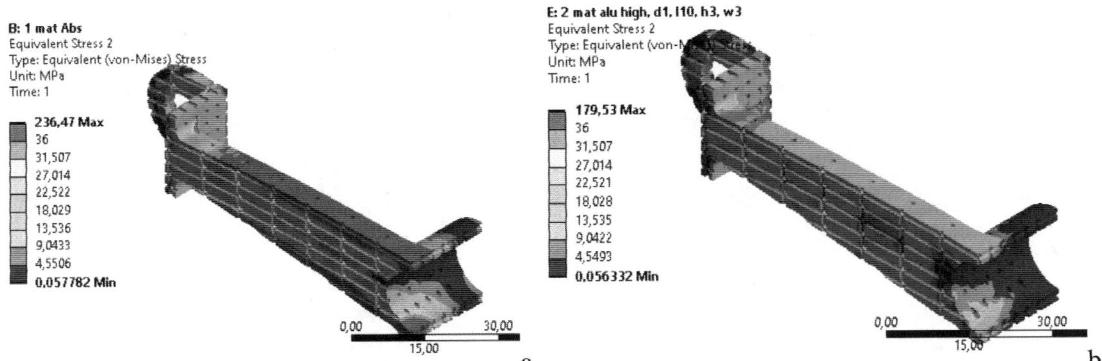

a b

Figure 11. Stress (Base):
a - ABS, without reinforcement,
b - ABS Base, with aluminum reinforcement

IOP Conf. Series: Materials Science and Engineering 1016 (2021) 012019 doi:10.1088/1757-899X/1016/1/012019

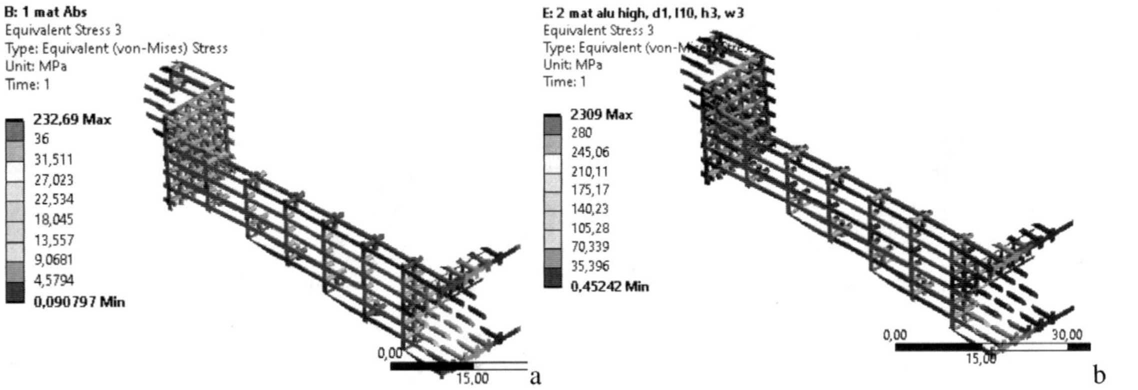

Figure 12. Stress Reinforcement:
a - ABS, without reinforcement,
b - ABS Base, with aluminum reinforcement

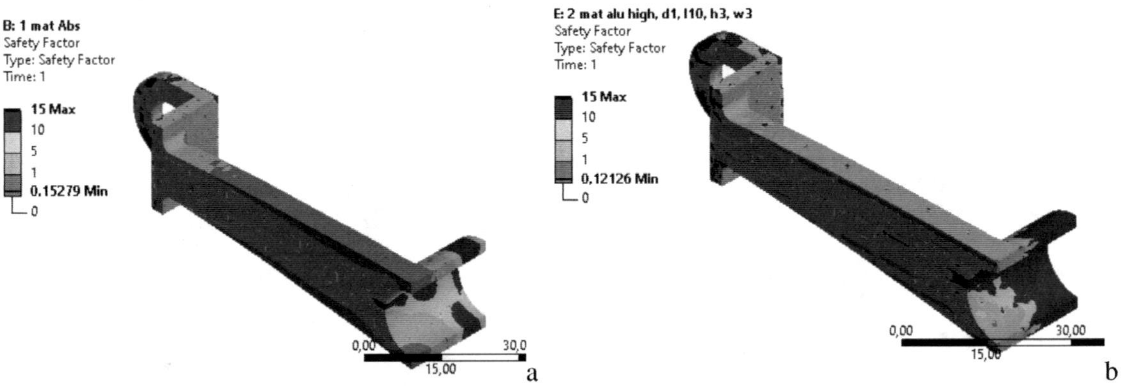

Figure 13. Safety Factor:
a - ABS, without reinforcement,
b - ABS Base, with aluminum reinforcement

Based on the research, a summary table of the results of the maximum stresses arising in the base material and reinforcement (see table 3).

Table 3. Maximum stresses that arise in the base material and reinforcement

		Base			Reinforcement		
	Total deformation (mm)	Hot spot Stress HS (MPa)	Maximum General Stress GS(MPa)	Safety Factor (SF)	Hot spot Stress HS (MPa)	Maximum General Stress GS(MPa)	Safety Factor (SF)
B	10,0	236,5	60	0,7	232,7	60	0,7
C	0,34	241	47	6,3	241	46,7	6,3
D	0,12	237,9	59,2	4,2	244,4	59,6	4,2
E	3,71	179,5	25	1,44	2309	540	0,52
F	2,17	218	14,1	2,57	3629,7	810	0,28
G	7,19	272	55	0,66	3928	940	0,3
H	2,0	127,7	15	2,4	1636,5	280	1
I	3,18	167	9,9	3,6	1330	401	0,7
J	2,0	127,7	16,7	2,15	1636,5	310	0,9

As we can see from the received research results of the interconnection between materials of a 3D model and geometrical parameters of a reinforcement frame, it was established that using a base model from ABS plastic, hot spot stress does not surpass 250 MPa, however, using ABS plastic with aluminum reinforcement, hot spot stress rises to 3928 MPa.

Conclusions

The use of computer-aided design systems and topological optimization modules makes it possible at the initial stage of modeling and development of parts to obtain their optimal parameters in accordance with the specified production conditions. Strengthening the structural parameters of parts made of composite materials by reinforcement, methods of generative design helps to increase their fatigue strength characteristics in places of greatest stress concentration.

Modern programs of topological optimization at this stage do not allow to optimize the parameters of parts of complex shape by reinforcement in the form of a regular structure with a combination of several materials. However, the process of parametric analysis and synthesis of bodies with a reinforcing mesh in the form of one of the regular structures can still be performed.

Such a structure can create problems for automatic optimization, because changing the mesh parameters can change the number of bodies, as well as the configuration of the faces to which the boundary conditions are applied. Arbitrary exit of the reinforcement mesh outside to the surface of the part creates the possibility of the appearance of local concentrators (hot spots). The stress values in such concentrators are random values and in a linear formulation can reach huge value (which will not happen in reality due to the possibility of plastic deformation). Thus, when evaluating the solution to the optimization problem, it is recommended to evaluate not the level of maximum stresses, but the stresses at the control points of the structure and the general stress, which do not surpass 60 MPa for a basic model from ABS plastic. Using ABS plastic with aluminum alloy AlSi10Mg reinforcement stresses rise to 940 MPa. For these materials, the safety factor is 0,3 та 0,28. Maximum stresses for steel 316L are only 810 MPa, so they are a bit lower in comparison with aluminum alloy for the same dimensional parameters of the reinforcing mesh.

Since the melting point of steel is 1440 °C and of aluminum - 590°C, it is possible to use the latter as a reinforcing material in 3D printing technologies on state-of-the-art high temperature 3D printers.

References

[1] Jaworski N Iwaniec M and Lobur M 2019 Implementation features of composite materials effective mechanical characteristics finding method based on microlevel cellular structural models *15th International Conference on the Experience of Designing and Application of CAD Systems (CADSM)* 1-5 doi: 10.1109/CADSM.2019.8779273

[2] Jaworski N Lobur M and Iwaniec M 2018 Implementation features of cellular composites microlevel structural models construction based on Voronoi tessellation and OpenCL technology usage *14th International Conference on Perspective Technologies and Methods in MEMS Design (MEMSTECH)* 109-13 doi: 10.1109/MEMSTECH.2018.8365713

[3] Pobereyko S Holovatyy A Sokolovskyy Y and Pobereyko B 2018 Mathematical Model and Software for Determination of Short - Term Strength of Anisotropic Composite Materials *IEEE 13th International Scientific and Technical Conference on Computer Sciences and Information Technologies (CSIT)* 207-10 doi: 10.1109/STC-CSIT.2018.8526639

[4] www.soel.ru/online/kniga-industriya-4-0-umnye-tekhnologii-klyuchevoy-element-v-promyshlennoy-konkurentsii-chast-4-simul/

[5] Dede L Borden M J and Hughes T J R 2012 Isogeometric Analysis for Topology Optimization with a Phase Field Model *Archives of Computational Methods in Engineering* **19**(3): 427-65

[6] www.autodesk.com

[7] Sika R Szajewski D Hajkowski J and Popielarski P 2019Application of instance-based learning for cast iron casting defects prediction *Manag. Prod. Eng. Rev.* **10** 101-7

[8] Kaluza A Kleemann S Frohlich T Herrmann C and Vietor T 2017 Concurrent design & life cycle engineering in automotive lightweight component development *in 1st Cirp Conference on Composite Materials Parts Manufacturing* **66** 16-21

[9] Tkachuk M A Bondarenko M Grabovskiy A Sheychenko R Graborov R Posohov V Lunyov E Nabokov A and Vasiliev A 2018 Thin-Walled Structures: Analysis of the Stressed-Strained State and Parameter Validation *Eastern-European Journal of Enterprise Technologies* **1/7** (91) 18-29

[10] Stolarski T Nakasone Y and Yoshimoto S 2018 *Engineering Analysis with ANSYS Software 2nd Edition* (Butterworth-Heinemann) p 553

[11] Zdobytskyi A Lobur M Iwaniec M Breznitskyi V 2019 Optimization of the Structural Characteristics of the Robotic System Holder *15th International Conference on the Experience of Designing and Application of CAD Systems (CADSM)* 1-4

[12] Zienkiewicz O C Taylor R L and Zhu J Z 2013 The Finite Element Method: Its Basis and Fundamentals 7th Edition (Butterworth-Heinemann) p 756

[13] Farmaga I Lobur M Shmigelskyi P Javorskyi N and Śpiewak P 2012 Regular and adaptive meshing algorithms for modeling of spherical inclusions by finite element method *Modern Problems of Radio Engineering, Telecommunications and Computer Science - Proceedings of the 11th International Conference (TCSET'2012)* 505-507

[14] Jaworski N Andrushchak N Lobur M Iwaniec M and Sitek W 2019 Composite material microlevel cellular model data transfer and verification by fem analysis software systems *IEEE 15th International Conference on the Perspective Technologies and Methods in MEMS Design (MEMSTECH)* 17-22 doi: 10.1109/MEMSTECH.2019.8817376

[15] Eschenauer H A and Olhoff N 2001 Topology Optimization of Continuum *Structures Journal of Applied Mechanics Reviews* **54**(4) 372

IOP Conf. Series: Materials Science and Engineering 1016 (2021) 012020 doi:10.1088/1757-899X/1016/1/012020

The Use of Information Systems for Regulation of Gas Engine Operating Parameters

A Kalwar[1], F Kurdziel[1], K Pytel[2]*, S Gumula[2], M Lobur[3] and M Melnyk[3]

[1] PGNiG Termika Energetyka Przemysłowa S.A. Rybnicka 6C, Jastrzębie Zdrój, Poland
[2] AGH University of Science and Technology, al. A. Mickiewicza 30, Krakow, Poland
[3] CAD Systems Department, Lviv Polytechnic National University,
Lviv, 5 Mytropolyta Andreya, 79013, Ukraine

*Email: kpytel@agh.edu.pl

Abstract. The purpose of the work is to indicate the possibility of using selected IT systems to analyse work and regulate operating parameters of gas engines. An additional task of the work is to indicate new possibilities of using IT systems in power plants. The publication shows the development of industrial information systems in recent years. It is shown how these systems are used for direct control and adjustment of operational parameters of power equipment, ranging from water boilers to steam boilers, turbines and cogeneration aggregates. These systems implement technical safeguards and can be used to analyse work characteristics in an appropriate configuration. The more elaborate energy systems, the wider the scope of use of IT systems. Due to the full automation and control of cogeneration gas-powered power generators, an automatic heating water flow control system through heat exchangers can be used in these installations. This solution requires the expansion of the heating water pumping system with pumping sets with inverters and fittings of the pumping station with remote control.

1. Introduction

Nowadays, as never before, the greatest emphasis is placed on efficiency, from the effective use of working time, through the efficiency of product and goods production to the efficient use of fuels. More specifically, this issue for industry is about the highest possible profit with the least effort while maintaining quality requirements. In the case of power plants, the raw material for the production must be used with the smallest profligacy; the production process must be carried out with the lowest possible consumption of raw material and energy while maintaining certain characteristics and quality requirements of the final product. With the development of technology, you can observe the rapid development of the IT industry. Industry has become a real training ground for using technical innovations in mass production. Engineers have increasingly begun to reach for IT solutions and use them for their purposes. Considering IT systems in the operation of energy devices, we can say that we are talking about a instrumentation and automated control systems most often in a closed system. Closed systems exist as preanalytical, analytical or post-analytical phase of the production cycle [1]. The supplementary hardware and software have been specifically designed to integrate with instrumentation and applications [2]. The control in the case of gas engines is the programmable control in the part covering only the gas engine, while the heat recovery system works in the programmable control mode with some elements operating in the fixed-value and follow-up control mode. The control is carried out in a local and a remote mode [3, 4]. The user decides about the mode control. When analysing the possibilities of using IT systems, it is necessary to take into account:

- the possibilities offered by the IT layer of this system, i.e. the capabilities of the system for entering measurements, connections with actuators, the possibility of modifying and expanding the system with further parts - these possibilities are important for specialists in the I&C industry
- the appearance of the user interface, whether it is readable or easy to learn the knowledge of plant control associated with the system; this is particularly important in the context of people who deal with the operation of installations and not the operation of IT systems
- the IT system should be secured against losing operator control over the technological process in any emergency
- the IT system should facilitate the operator's work.

The quality of an IT solution is a combination of technical, ergonomic and functional quality. The visualizations of the implemented tasks is very important from the point of view of the practical operation of I & C equipment. The publication compares the readability of information obtained in textual form with information resulting from the process of visualization [5, 6]. The aim of the analyses is to point IT solutions that take into account the importance of functional and ergonomic factors. The simplicity of learning, an effortlessness of use, a visual mapping accuracy, shorter system operation training, faster execution of tasks, and a reliability of system use determine the success of control, measuring equipment and automation systems. After analysis of the technical condition of the existing configuration of I & C equipment, a response was prepared that provided a solution to the illegibility of existing solutions. The system was controlled by an automation system controlled by the superior ASIX SCADA system. The presented solutions are intended to show the possible user interface modifications that result in increased readability of the installation's work. The basis of the work is to present the graphic representation of combustion process parameters in the existing gas engine.

2. A gas engine as a high-efficiency cogeneration unit

A heat engine is a machine that converts a heat energy obtained by burning solid, liquid or gaseous fuels into a kinetic energy. The discussed engines are internal combustion engines converting a thermal energy obtained as a result of combustion of gaseous fuels, in which the conversion of the thermal energy into the kinetic energy takes place inside the engine working space (in the cylinder). The engines are made in the technology of vertically arranged cylinders, with a turbocharging (one turbocharger for each cylinder row).

The motivation for achieving high values of fuel utilization factor has led to increased interest in cogeneration applications [7]. The analysed gas engines coupled with electric generators with electric power from 2.0 to 4.3 MWe. These engines produce electricity and the heat from the engine block and the heat contained in the exhaust gas is recovered as waste and utilized in further processes. Thermal energy is transferred to the heating network and used for the production of cold. The first application of the analysed internal combustion engines together with an electric generator was the replacement of previously used gas boilers powered by mine gas, with a capacity of about 5.0 MWt that was used to meet the needs of the mine-heating network. The use of gas engines means that with the same amount of gas fuel consumed, heat energy is obtained and electricity is generated as an additive.

After these positive experiences with replacing small water boilers with gas engines equipped with generators, it was decided to use the heat generated by gas engines to generate cooling energy from this heat. Due to these investments, installations containing gas engines began to be increasingly perceived as modern and energy-saving energy installations. Small gas boilers were simple installations and required only a few basic parameters to be monitored:

- a water flow through the boiler
- temperatures before and after the boiler
- a pressure in the combustion chamber
- a gas pressure
- a gas flow to the boiler.

Supervision over these parameters was carried out by observing the indications of instruments installed on control boards. The boiler protection was carried out by electrical systems, control and measuring instruments and mechanically executed. However, small gas boilers were fairly simple installations and required only a few basic parameters to be controlled, it became necessary to develop automation, control, and measurement when used for large power installations with gas engines. Matching tables with control and measuring instruments became larger and with many different indicators. Installations with the production of cooling energy from heat energy have become technically predominantly advanced. The control of aggregates with gas engines is a closed-loop control system with electric actuators. By requesting engine start, the engine operator starts the programmed engine start sequence, meeting all conditions causes the engine to start. When the engine is running, the operator sets the value of the engine load, and the system controls the engine cooling system so that the oil and cooling water temperature parameters are optimal [8]. Installations that support various functionality of gas engines have become almost ideal for setting up the new design of computer systems. Because these installations cooperate with various types of other external entities, it became necessary to connect various types of computer systems with various functionalities. Some systems control, supervise and implement security, others only visualize parameters, and others collect data and present them in the form of reports. It is a training ground for testing and development of systems that can integrate all functions [9].

3. A control of the gas engine system

3.1. Selected problems of the gas engine operation

Optimizing the operation of a gas engine, especially of high power, is an extremely important concern. The control system for the integration of automation and information systems, as well as production should be responsible for:

- an engine management
- an adjusting the engine cooling system on the coolant side
- an engine lubrication control.

Controlling engine operation results from the need to obtain the assumed system efficiency. The use of this system functionality eliminates the risk of starting the engine with out of order engine systems. When starting the engine, the operator triggers a programmed start sequence during which many systems are tested:

- a throttle (the system checks whether it moves in the full range of operation - in the event of non-compliance interrupts the start procedure)
- a bypass valve of the turbocharger (in the same range as the throttle)
- the system checks if there are adequate pressures in individual elements of the cooling system
- after the start-up, the system ensures the correct ignition setting of the engine, as well as the right mixture temperature.

In addition, the system causes the engine to be shut down if the limit parameters are exceeded. Adjustment of the engine cooling system on the coolant side is the use of an IT system that facilitates the operation of the engine, the operator corrects only to a small extent the settings of the engine cooling system, the system reacts to any changes in the temperature of the coolant in accordance with programmed algorithms. In the event of a sudden increase in the temperature of the coolant due to the deactivation of a heat exchanger, the emergency coolers are started in the correct order at the appropriate speed of the cooling fans until the load on the engine is reduced or it is taken out of service. The system prevents unnecessary engine starts, and thus increases the reliability and efficiency of production.

Adjustment of the engine lubrication system is a system that decides whether in the event of an oil level dropping in the oil pan top up its level by automatically pumping from the fresh oil tank. The oil temperature is one of the basic parameters necessary for the trouble-free operation of the gas engine and this parameter is also one of the most closely monitored parameters. The computer system immediately responds to any temperature changes by continuous analysis of the thermal state of the engine. It does faster than the operator by reacting at many points in the installation simultaneously, which is not possible with manual adjustment of the system by the operator.

3.2. The gas engine control solutions
The first and the basic IT system is the gas engine control system. This system allows the gas engine and the generator to be operated safely and reliably. To supervise and control the operation of gas engines, it uses an IT system that provides the ability to control the engine from the place of engine installation locally as well as remotely.

Figure 1 presents a simplified diagram of connections and cooperation of computer systems of the cogeneration aggregate and connections with external measuring elements. The basic system collects data from the gas engine measuring points:

- a coolant temperature
- an oil temperature
- a lubricating oil pressure
- pressure measurements in the cooling system
- a combustion temperature and an ignition energy in cylinders.

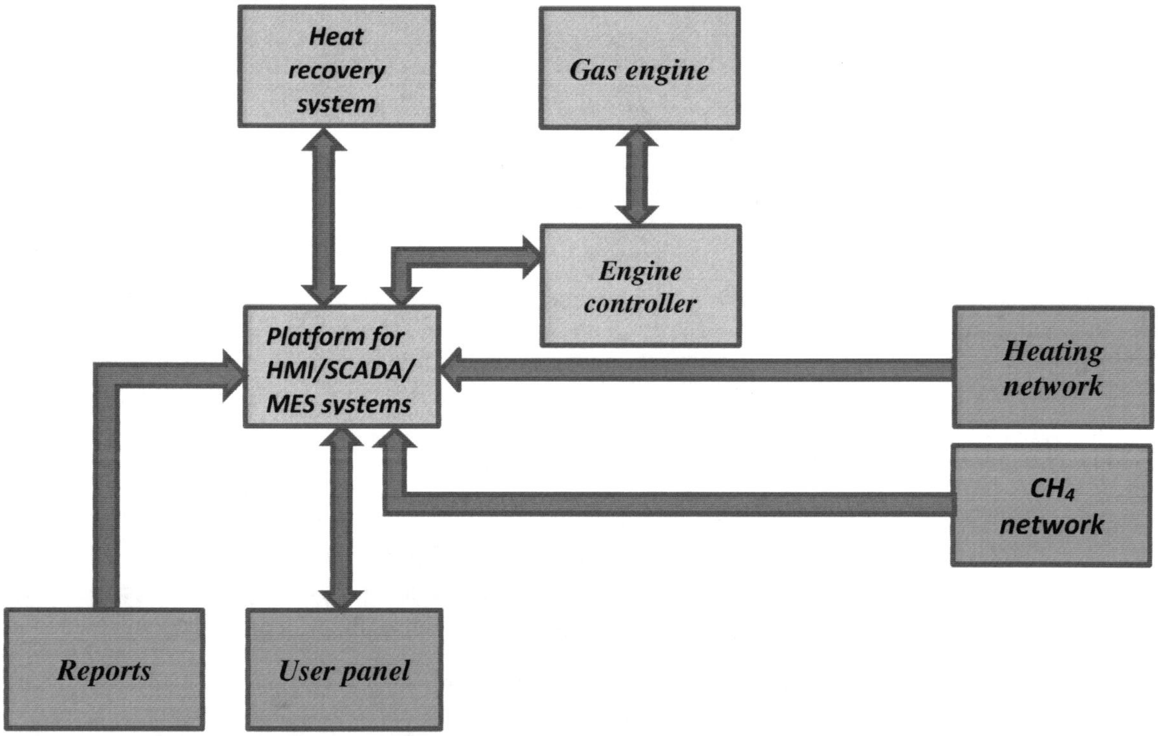

Figure 1. The trigeneration installation diagram.

Control of gas engines is carried out individually for each of the engines. Each engine is located in an individual building. In the building of each of the engines on the control cabinet there is a tactile control panel of the TEM Evo system and a synchronoscope ignition. In the room of the block control room of the CHP plant, there are computers with an access application to the TEM Evo system installed - the access management to the TEM Evo system is carried out from these computers. In the place of the block heat and power plant control room there are computers with ASIX application with full access to control and regulation of gas engines, as well. In the engineering and technical service rooms there are computers with ASIX application installed, but only with the possibility of viewing operating parameters. The SCADA system is the master system is ASIX system of ASKOM Gliwice. To control the processes in gas engines, SIMATIC controllers from Siemens are used:

IOP Conf. Series: Materials Science and Engineering 1016 (2021) 012020 doi:10.1088/1757-899X/1016/1/012020

- SIMATIC S7-300 for TCG 2032 V16 engine
- SIMATIC ET 200S for CG 260 16V engine

The standard SIMATIC controllers consist of:

- Central Unit - CPU
- Expansion modules - with a central cassette (CC) with a capacity of up to 8 modules, in the case of more modules, expansion cassettes (EU) are used. A maximum of 32 modules can be supported (4 cassettes of 8 modules each)
- Interface modules - responsible for communication between individual cassettes
- Function modules - have processors and work independently of the controller's main processor. They are used when technological tasks require high computing speeds and high control dynamics
- Communication modules - allow the controller to be connected to a network such as PROFIBUS.

The WOODWARD GW4 controller is responsible for the process of automatic synchronization with the power network in both motors. The SCADA system collects data from the SIMATIC controller of the relevant engine, using these data it partially realizes control algorithms, on the basis of available data it calculates other data which are not measured. The system issues alarm and warning messages, archives data, gives the possibility to create archive charts from any time horizon.

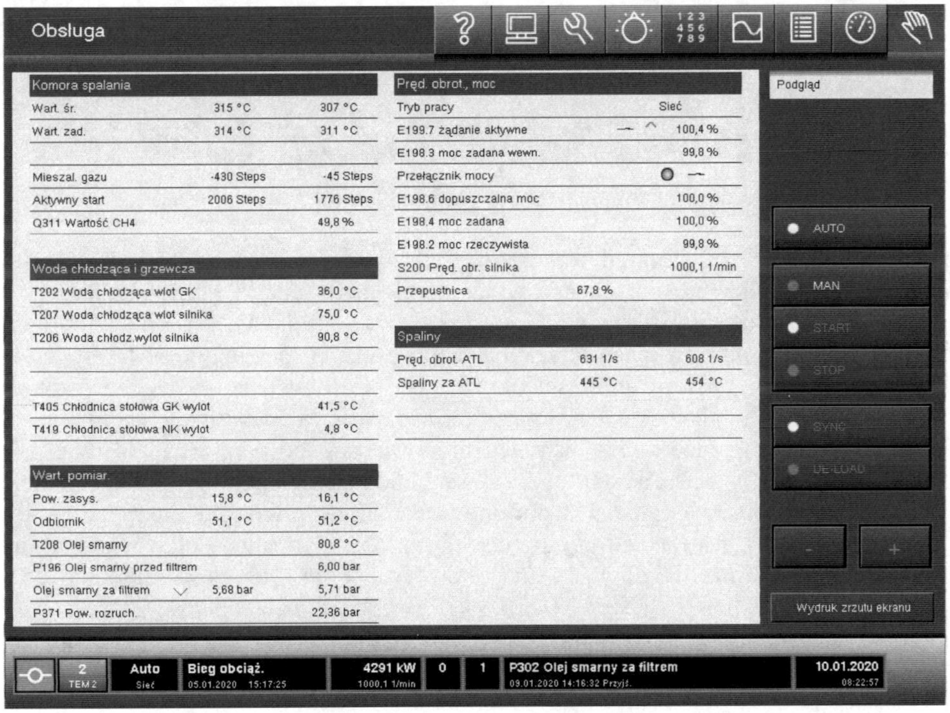

Figure 2. The real cogeneration aggregate control display.

The Figure 2 shows the previously used by the authors and the currently existing user interface fully functional but not very legible. The figure shows the TEM Evo system user interface, the basic hood from which the engine starts and stops, as well as its synchronisation with the power grid and the possibility to change load. Moreover, this display selects the engine control method - manual or automatic, additional basic information related to the gas engine operation is provided on this display. It is a fully functional display, however, it is not very readable, which causes some difficulties for people operating the installations, additionally it has very limited functions of analysing the operating parameters of the archives. This causes some difficulties for people who operate installations.

IOP Conf. Series: Materials Science and Engineering 1016 (2021) 012020 doi:10.1088/1757-899X/1016/1/012020

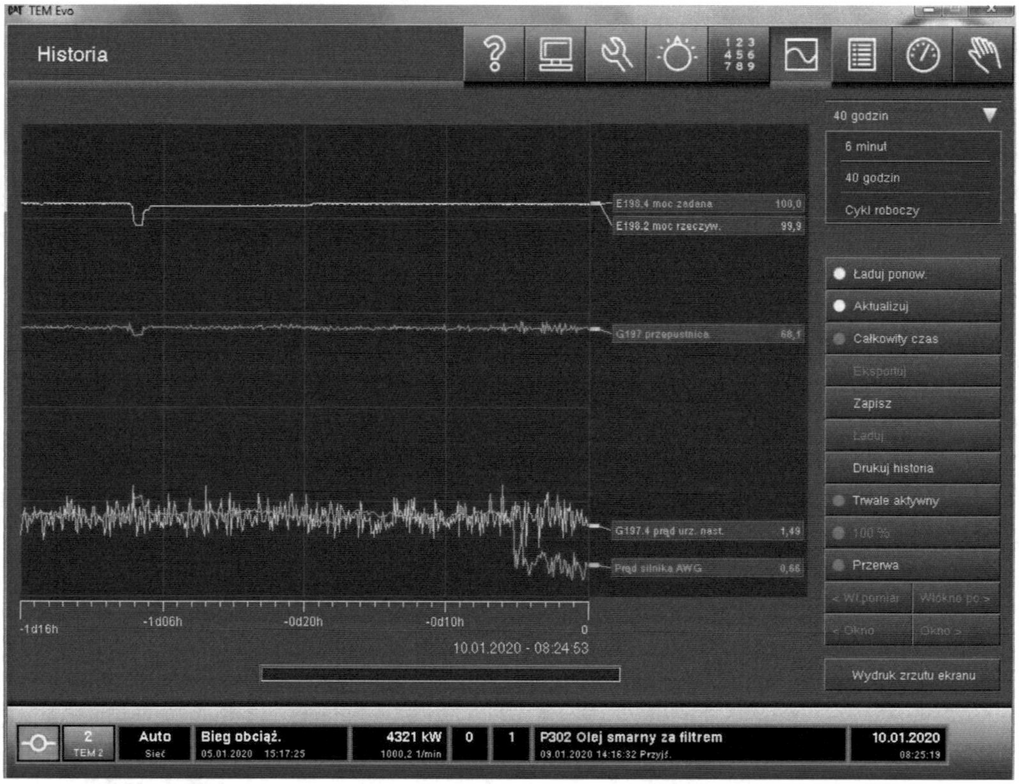

Figure 3. The real analogue measurements and trend display.

The presented interface has limited functions of analysing archive operational parameters, as shown in the Figure 3. The function that draws archival charts allows you to generate charts n 6 minute or 40 hours back, but without the ability to save charts, except for the option to save to external media. In addition, the number of parameters that can be placed on one chart at the same time is limited. The 6-minute time horizon can be selected on this figure. It is not possible to stop it - only a screenshot to *.pdf format can be taken. The chart can display up to 20 parameters simultaneously, but then the chart becomes unreadable. The attached graph shows the parameters of engine operation allowing to check its load capacity: actual load, set load and throttle opening degree together with the current of throttle flap control. As for the opportunity of interfering with the engine operation via the interface, it is possible to select the operating mode (automatic or manual), start and stop the engine, synchronize and unsynchronize the engine with the power grid, as well as increase and decrease the engine load.

The Figure 4 presents a text system informing the operator about changes occurring at a given moment. The basic operating parameters of a gas engine are shown only in the form of numbers. It is not possible to view changes, trends or select individual parameters for analysis.

The Figure 5 graphically illustrates the basic operating parameters of a gas engine. This shows a typical control display, which was created in cooperation between operators and designers. The aim of this cooperation was to create such a display of the technological system that would transmit the necessary information while remaining as legible as possible - this task was fully achieved. This is the extended equivalent of the display shown in Figure 4, which uses the available data acquisition and interpretation possibilities. This display in a simple and legible way shows individual parameters with the possibility of viewing graphical charts of changes of individual parameters, this functionality allows for thorough analysis of the installation's operation and interpretation of changes in its parameters.

The discussed graphic representation of combustion process parameters in the engine informs and controls the operation of the gas engine and supervises the basic parameters of its operation.

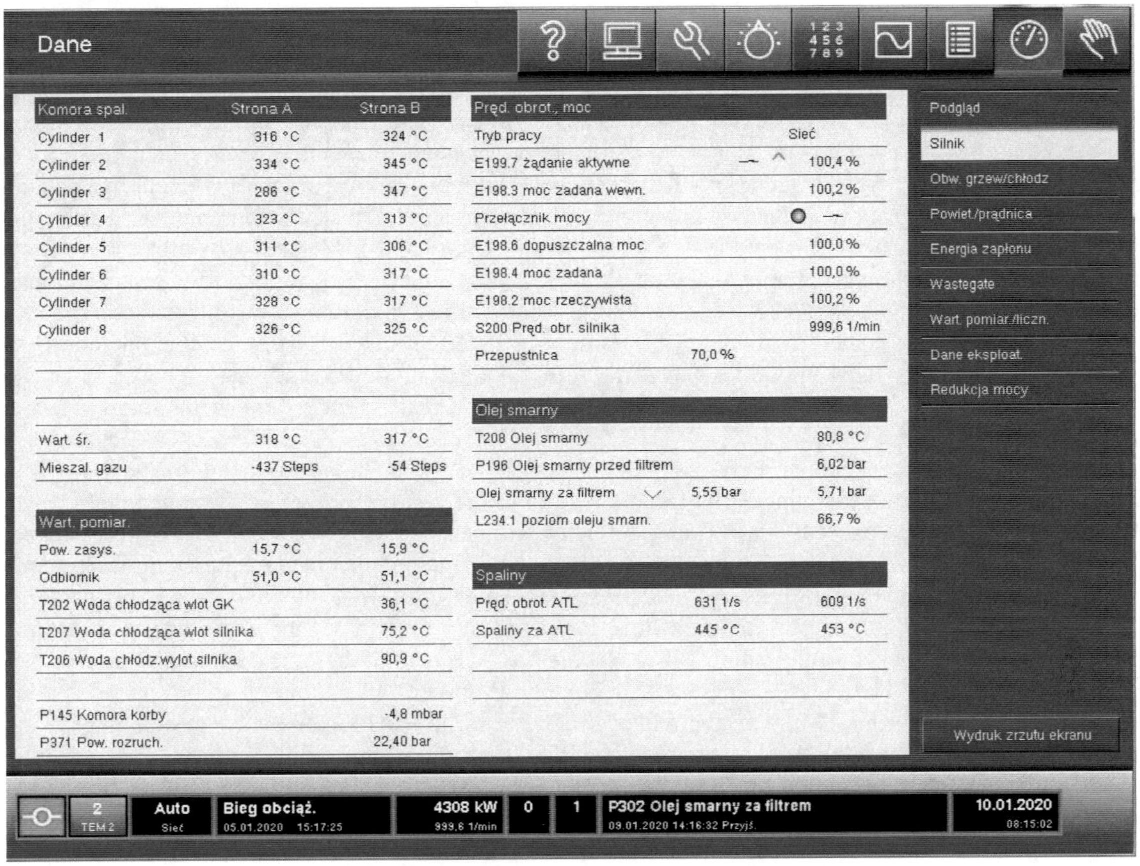

Figure 4. The real text presentation of the combustion process parameters in the engine display.

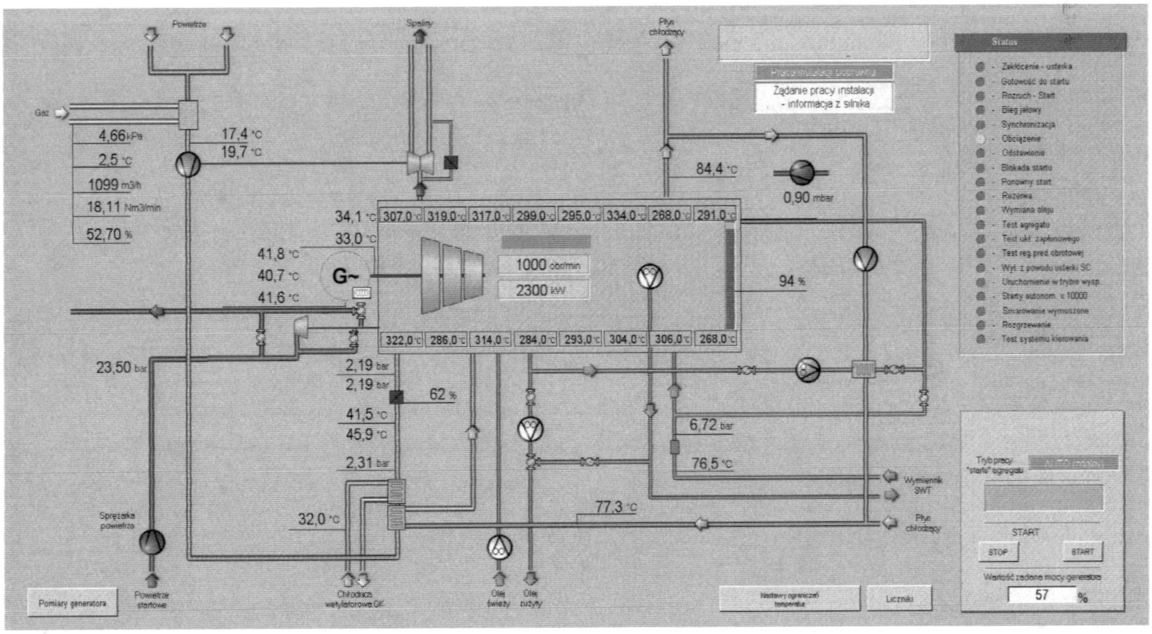

Figure 5. The real graphic representation of combustion process parameters in the engine display.

Conclusions

This article presents solutions using IT systems for the purposes of supervision and control of exploitation, as well as technical analysis of high-efficiency cogeneration aggregates using gas as the fuel. Analysing the project of processes visualization it can be argued that the legitimate goal has been achieved, but the process of operating cogeneration aggregates can be even more automated.

Even the simplest energy installations currently use support from computer measurement and automation systems. The use of the computer systems allows a very sensitive industrial user to change parameters to ensure the safe operation of these installations, and gives the possibility of remote control of power station installations. In the location of the computer station with the appropriate control software installed, it is possible to select the place of the installation control interface. The use of IT systems in the power industry enables ongoing detailed supervision over the operation of devices. Computer systems support the operator in conducting operations that implement many adjustment processes outside the operator. This approach protects devices and installations against damage.

Unwanted consequences of such a solution can certainly include the need to have highly specialized knowledge among those involved in servicing the control system of these installations. It happens in practise that exceeding a certain parameter will cause the installation to be shut down, and the reason is not a physical exceeding of the measured value, but a defect of the measuring instrument or the measuring path. In such cases, it is necessary to have knowledge of how the control cycle is implemented and what are the relationships between the individual measurement parameters. Implementation of the control system allows achieving effects in the form of improved system efficiency and more efficient system management. Prepared visualization contributes to the identification of the system failure and provides all necessary information about the system operation. The system of visualisation need to be expanded, so this is not a final solution.

References

[1] Kenett R S Ruggeri and Faltin F W 2018 *Analytic Methods in Systems and Software Testing* (New York: Wiley–Interscience)
[2] Kenett R S Zacks S and Amberti D 2013 *Modern Industrial Statistics: with applications in R, MINITAB and JMP (Statistics in Practice)* (New York: Wiley–Interscience)
[3] Legierski T Wyrwał J Kasprzyk J and Hajda J 1998 *Programowanie sterowników PLC* (WPKJS)
[4] Kwaśniewski J 2005 *Programowalne Sterowniki Przemysłowe w Systemach Sterowania* (ZP Roma-Pol)
[5] Sacha K 1998 *Sieci miejscowe PROFIBUS* (MIKOM)
[6] Seta Z 2002 *Wprowadzenie do zagadnień sterowania: wykorzystanie programowalnych sterowników logicznych PLC* (Mikom)
[7] Ancona M A Bianchi M Branchini L and Melino F 2014 *Energy Procedia* **45** pp 1225-1234
[8] Pytel K Suchanicz J Livinsh M and Sternberg A 2013 *Condens. Matter Phys.* **16(3)** 31706
[9] Prauzner T and Ptak P 2011 *Journal of Technology and Information Education* **1** pp 292–296

Acknowledgments

The project is financed by the Polish National Agency for Academic Exchange, No. PPI/APM/2018/1/00049/U/001.

IOP Conf. Series: Materials Science and Engineering 1016 (2021) 012021 doi:10.1088/1757-899X/1016/1/012021

Analysis of the Possibilities of Using Automatic Control Systems for Operating Gas Engines for a Chilled Water Production

F Kurdziel[1], A Kalwar[1], K Pytel[2]*, S Gumula[2], U Marikutsa[3] and D Korpyljov[3]

[1] PGNiG Termika Energetyka Przemysłowa S.A. Rybnicka 6C, Jastrzębie Zdrój, Poland
[2] AGH University of Science and Technology, al. A. Mickiewicza 30, Krakow, Poland
[3] CAD Systems Department, Lviv Polytechnic National University,
Lviv, 5 Mytropolyta Andreya, 79013, Ukraine

*Email: kpytel@agh.edu.pl

Abstract. The publication presents the impact of using automatic control systems for operating gas engines aimed at a chilled water production. The purpose of the work is to analyse the possibilities of using automation systems to regulate gas flow. An important parameter of fuel supplied to gas engines used in the production of cold is its amount, which depends on the pressure. Unstable operation of gas distribution systems caused by an extensive network, many customers and operation of devices working with different loads causes improper, unpredictable operation of other devices working in the network. Uncertainty of fuel parameters results in unstable operation of gas engines, their load, and variable production of electricity, heat and cold. Ensuring proper and stable production is a key element of production processes. The concept of building a gas pressure regulation system based on its amount in the network will ensure the required, stable operation of gas engines and refrigeration units coupled with them, and thus provide a guarantee of production of generated media. The presented solution provides a solution to existing problems. The system is based on two dampers coupled to each other and controlled by an automation system controlled by a superior SCADA system.

1. Introduction

Energy obtained from the power system comes from a number of sources. Energy in various forms is converted into electricity in generators. The cogeneration unit is formed from the expansion of the generating set with an element for heat recovery, which leads to the use of primary energy contained in the fuel for the production of electricity and heat. Cogeneration is the production of heat and electricity in the most efficient way, because in one technological process. It is characterized by greater energy efficiency of the system by up to 30 percent than in the case of separate production of electricity and heat [1]. Expansion of cogeneration units with additional installations allows optimizing energy use and creates a polygeneration solution that allows better use of energy. Trigeneration systems are used in the medium and high power range systems [2]. Trigeneration is the technologically associated production of thermal, mechanical (or electrical) energy and utility generated chilled water, aimed at reducing the amount and cost of primary energy necessary to produce each of these forms of energy separately. In heating systems, in the summer, it improves the cost-effectiveness of electricity production in combination with low heat demand and existing demand for chilled water [3]. The fuel used to power the polygeneration systems may be gaseous fuels. The main problem is ensuring acceptable gas quality by the internal combustion engine. One of the features of coking coal deposits, which comes from the mine, is the release of large amounts of methane gas (CH_4) during mechanical mining of coal seams. This causes a serious threat to safety for miners working underground due to the explosion of this gas, which occurs in the range of 4.5% to 15% of the mixture with air. To eliminate this threat, the mines have extensive methane drainage systems, which include underground methane drainage pipeline network and surface methane drainage stations [4 – 6]. The role of the methane drainage station is to extract methane from the bottom of the mine, and then compress it to a pressure enabling its injection and transport by gas network to the heat and power plant. From the beginning of operation, the power units of the plant have been prepared for methane combustion. Gas burners have been installed on steam boilers of these plants for both coal ignition and continuous combustion of methane as an additional fuel. The power of the burners was matched to maximize the use of available methane amount. The next stage of work was the construction of power generators based on piston gas engines aimed at the most effective use of methane. The analysed installation currently operates 7 gas engines.

The aim of the research is to determine a correct and stable process of a gas pressure regulation system. After careful analysis of the technical condition of the existing configuration, a response was prepared that provided a solution to existing problems. The system was based on two dampers coupled to each other and controlled by an automation system controlled by the superior SCADA system. The carried out studies show the effect of constructional and exploitation efforts to increase the efficiency of the chilled water production installation.

2. The issue of chilled water production

Chilled water installations are used as cold water distribution systems and are used in chemical and process engineering. Cooling installations working for the needs of chilled water production are selected depending on the availability of energy carriers, which are heat or electricity. An important feature of cooling installations based on chilled water is the high cost of building the installation, which is also associated with an appropriate control and automation system, and relatively low operating costs. This is the reason why chilled water installations are installed in combined heat and power plants. In chilled water systems, the distribution of cold water to heat exchangers is done by circulating pumps and pipelines connected to the chilled water aggregate. Chilled water produced in combined heat and power plants is generally used in nearby industrial facilities carrying out technological processes where there is a demand for cold. Chilled water power plant systems can operate on the basis of compressor, absorption or jet systems. Absorption systems are powered by thermal energy. They enable chilled water to be obtained at a relatively low supply temperature. These systems are reliable enough and do not need to be steam powered [7].

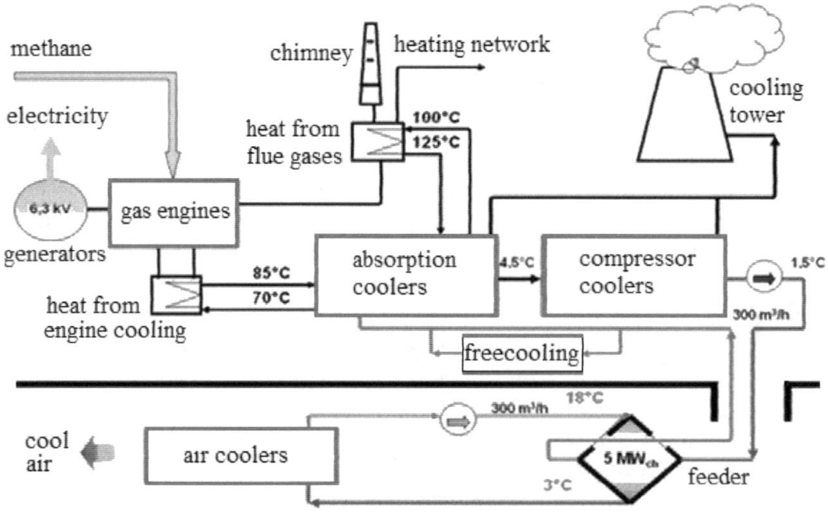

Figure 1. The trigeneration installation diagram.

IOP Conf. Series: Materials Science and Engineering 1016 (2021) 012021 doi:10.1088/1757-899X/1016/1/012021

The general concept of the project is to generate electricity using generators driven by lean methane piston gas engines, recovering heat from exhaust gases and engine bodies, transmitting cold using chilled water at a set temperature generated in absorption coolers powered by heat from engines and in compressor coolers powered by electricity from generators. Surpluses of electricity and heat are sold to the electricity and heating networks. Cold water is sent to the bottom of the mine and after reduction of static pressure, it is directed to air coolers, as shown in the Figure 1. The gas engine installation coupled with the cooling system uses methane gas as the fuel. The pressure range and gas concentration have a significant impact on the operation of the gas engine, its stability and performance. For purposes of the project, the range of correct pressure parameters for the gas engine accurate work is 4kPa - 30kPa while the concentration should be higher than 45%. The concentration smaller than 45% causes the UART (universal asynchronous receiver-transmitter) operation that results in unloading the engine until it stops completely. Due to the large diameter of gas pipelines (DN500) and the considerable distance from the methane intake station, the gas pressure is very unstable. Maintaining gas pressure at a level that ensures stable operation has become a problem, regardless of pressure fluctuations in the network.

3. Analysis of control and operation of gas regulation systems

The existing system consisted of two DN500 valves (shutoff valve was used for regulation, as well), as shown in the Figure 2. Its design and installation on such a large diameter allowed only a small pressure regulation within a range 5% of the valve diameter. Above this range, there were no changes resulting in a adjustment in pressure. A DN250 gate valve with AUMA electric drive with manual control was installed. The control was performed by the service (close / open) based on the pressure indication. This solution did not guarantee a quick response from the service. This affected the unstable operation of the systems until they were completely turned off. This resulted in production losses.

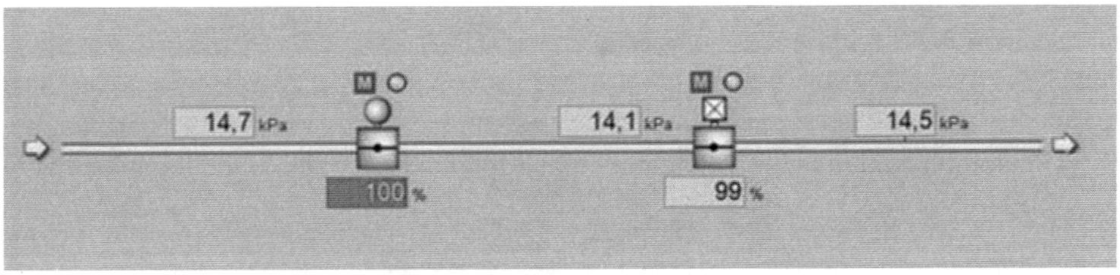

Figure 2. The gas regulation system diagram.

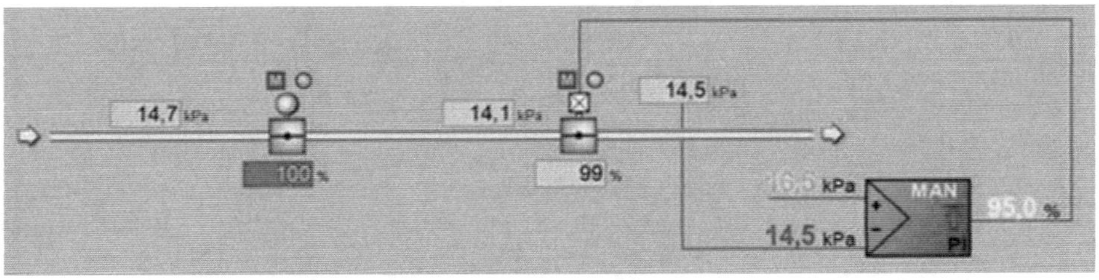

Figure 3. The gas pressure regulation diagram supplemented with a control element.

The design and implementation of an automatic gas regulation system based on and using existing devices and infrastructure was carried out. Industrial valves are used to control the resulting flow rate by opening or closing the valves. The concept of the new solution was based on the creation of an automatic control system that maintained a stable pressure. Two additional measurements were used for this purpose, as shown in the Figure 3. These were pressure measurements before and after the regulating valves. Data from analogue sensors and the position of the gate valve has been introduced to the superior control system based on the Siemens Simatic controller. Manual control has been preserved, which means that it can be controlled independently from the automation systems. The automatic control system was built based on the pressure measurement signal behind the throttle valve, as shown in the Figure 3. Setting the pressure in relation to the measured pressure measurement causes the throttle valve to open or close until the fixed pressure is reached. Determining the selected pressure is done by entering its value into the set pressure box in the control system (green in the Figure 3.).

Due to the structure of the system in which the control takes place, we can choose between control systems operating in the open and in the feedback system. The purpose of the open control system is to shape the output signal according to the program determined by the assumed signal. The assumed signal can be introduced by human, by measuring devices informing about the occurrence of a particular state of the system, or by devices that change signals according to the assumed program, as shown in the Figure 4. A characteristic feature of the open control system, both manual and automatic control systems, is the lack of information about output signal values and lack of guarantee of full control purpose achievement. These systems perform relatively simple functions and do not allow obtaining high control precision. Low accuracy of open control systems is associated with the occurrence of interference in the operation of the installation.

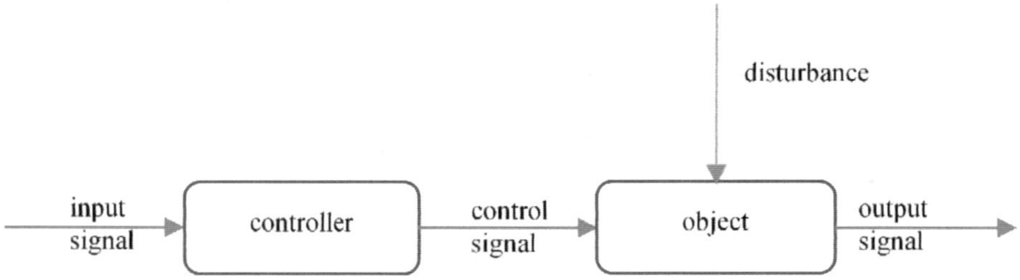

Figure 4. The general block diagram of the open loop system.

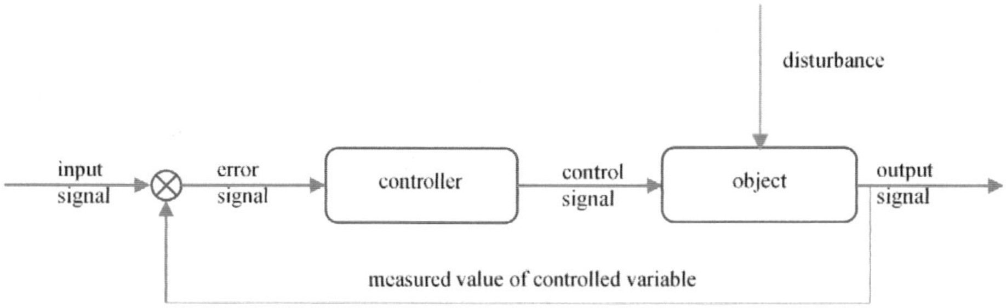

Figure 5. The general block diagram of a feedback control system.

The output value should be measured to compensate for the effect of interference on the output signal, then compare with the set value, and use the result of the comparison for corrective operation of the control device. Implementation of this action is carried out due to the existence of the negative feedback, as shown in the Figure 5. The advantages of using such a control system are that you can take corrective action that occurs regardless of the source and type of interference and there is no need to make a detailed process model. In addition, the versatility and reliability of the solution does not require complete knowledge about the process. When changing requirements, you only need to tune the controller for the proper functioning of the installation. The existence of the feedback loop is a characteristic feature of closed control systems and this resulted in the implementation of the solution into the project.

The new system's supervision and control system fulfils the following tasks:

- a visualization of the state of technological drives on mnemotechnical diagrams displayed on computer screens
- a drive control
- a visualization and archiving of analogue values
- archiving of alarm messages with the possibility of viewing and analysing
- provides native both historical and real-time trends and analysis using charts, as shown in the Figure 6; all the measurements are recorded by the system.

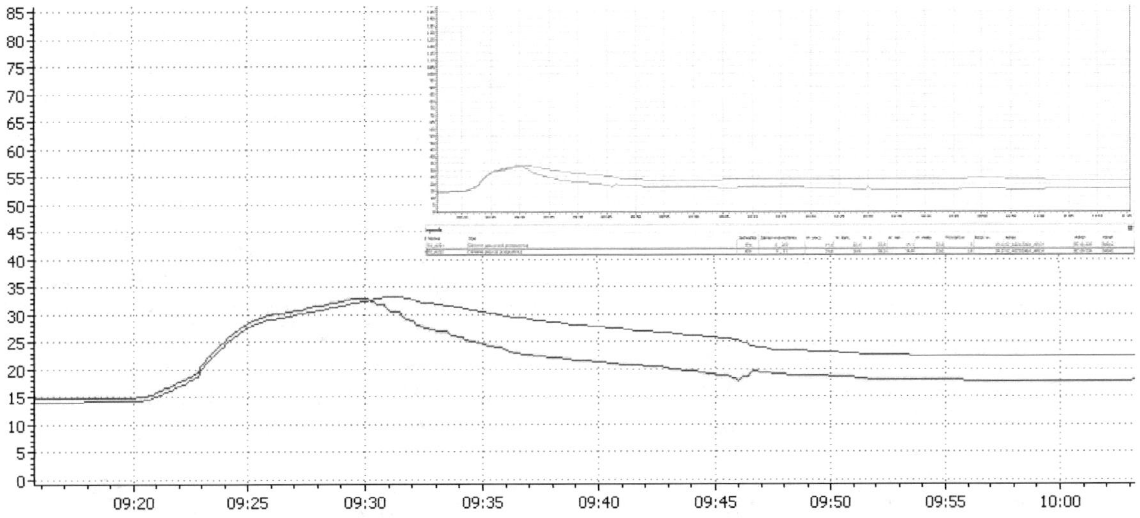

Figure 6. The visualization of graphs (trends) of analogue measurements.

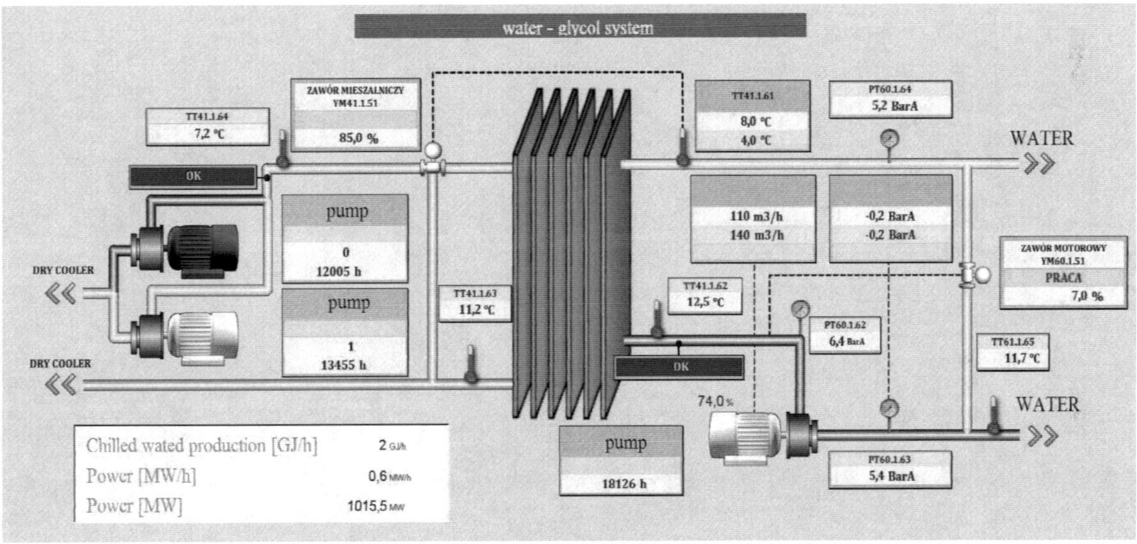

Figure 7. The graphic user interface display.

The graphical interface of prepared the ASIX SCADA system offer a plant representation spread over some levels. The graphic user interface, as shown in the Figure 7, includes:

- drive icons
- throttle / valves opening controls
- information about analogue measurements
- indicator for setting controller parameters
- graph of regulator.

Detailed views offer the possibility of observing several parameters. At this level, all the details can be displayed. The prepared system have several types of display demonstrating:

- a valve drive control
- a station view.

The throttle / valves opening control can be implemented on the drive control display. The control can be implemented immediately, as shown in the Figure 8, by '+' and '-' buttons and by the slider control on the display.

IOP Conf. Series: Materials Science and Engineering 1016 (2021) 012021 doi:10.1088/1757-899X/1016/1/012021

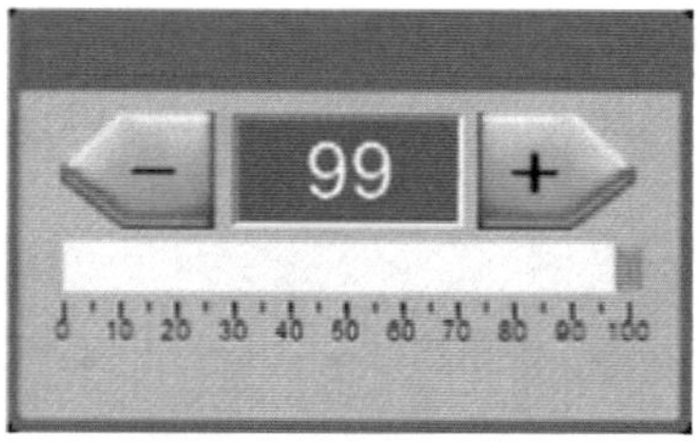

Figure 8. The drive control display.

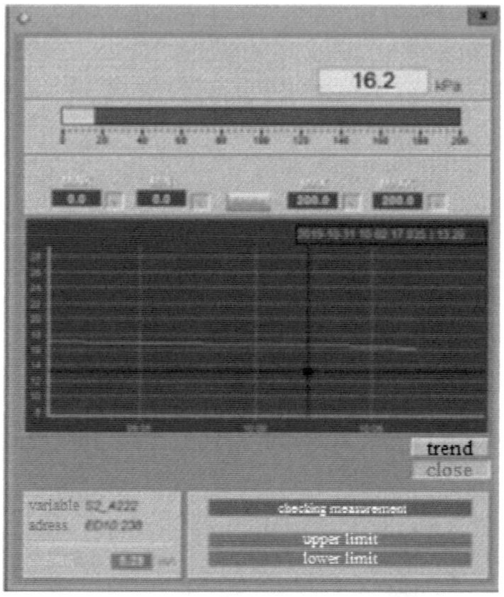

Figure 9. The analogue measurements and trend display.

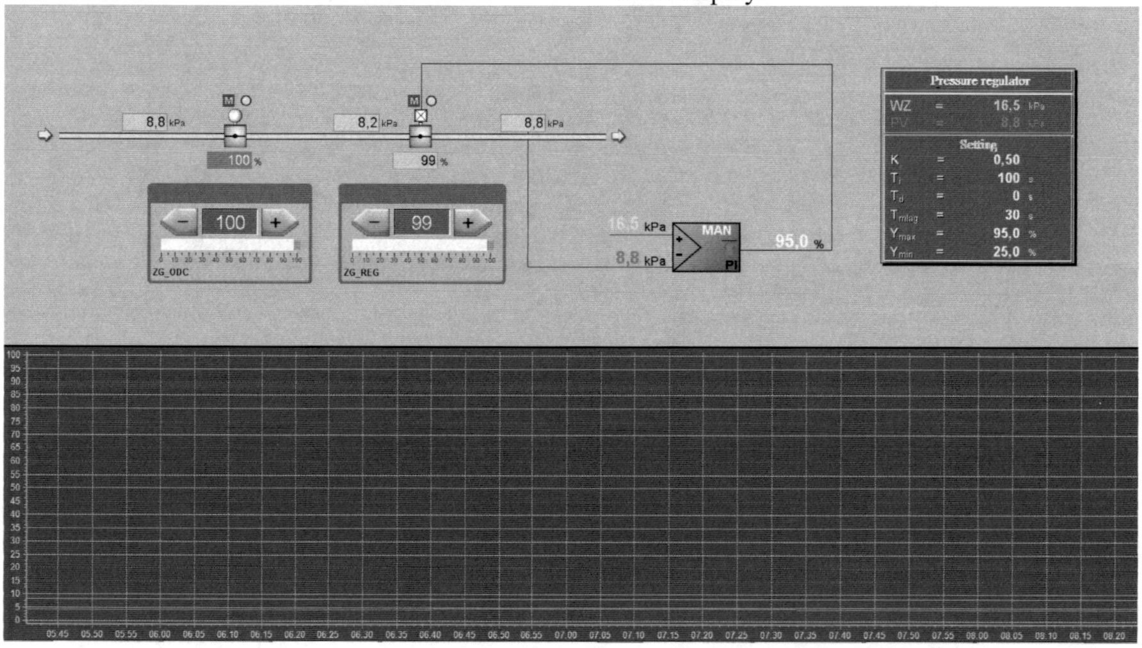

Figure 10. The information on gas quality and gas network operation status display.

The analogue measurement display, station view, can be opened to check the value of any analogue measurement. The display, as shown in the Figure 9, can provide the following information about a variable: description, unit, analogue value recorder, analogue value displayed as a number, analogue value in the form of a bar, display name and trend button provides native both real-time and historical measurement and trends. Modern devices are equipped with automation systems to facilitate work. HMI (Human Machine Interface) interfaces look like screens through which you can enter information into the system, as well as download and display production data.

The implementation of the regulation system, as shown in the Figure 10, has allowed the identification of problems and a simple solution to fluctuations in gas pressure supplied to gas engines [8]. As a final effect, stabilization of system operation was observed, which results in the stability of chilled water production [9].

Conclusions

This paper presents the use of the regulation system, the use of which for one element of the system, has resulted in significant benefits for the entire system for chilled water production obtained from the absorption heat pump working in the heat power plant system. The assembly and implementation of the described regulation system, in a simple way, allowed solving the problem of fluctuations in gas pressure supplied to gas engines, which stabilized the work of devices producing chilled water, resulting in work efficiency, as well. The implementation of the proposed solution resulted in the high predictability operation of the steam power plant equipped with an absorption system. Because of the solution used, the stability of gas pressure regulation for devices using it as fuel increased, which results in stable operation, operational reliability, reduction of failure frequency and occurrence of defects. An important aspect is also reducing the role of supervision by the staff over the important process of gas pressure regulation, which results in focusing the attention of the staff on other important elements of the process.

Most automation systems work in the assistance of man because human-machine interactions are important in current control systems, and automation system operators can affect production parameters. Implementation of the control system allows achieving effects in the form of improved system efficiency and more efficient system management. The implemented system provided an assistance in the analysis of the system work. Another conclusion could be a design solution that protects the unstable system operation [10]. This simple solution enabling the trouble-free operation of the system. In the analysed chilled water production system due to automation systems operating conditions, obtaining the required chilled water parameters has become more predictable. Implementation of the control system allows achieving effects in the form of improved system efficiency and more efficient system management [10]. The prepared system contributes to an increase in the reduction of system failure provides the necessary information about the operation of the system [11 - 13].

References

[1] Thumann A and Mehta D P 2013 *Handbook of Energy Engineering, Seventh Edition. 7th Edition* (Fairmont Press)
[2] Ganapathy V 2017 *Steam Generators and Waste Heat Boilers: For Process and Plant Engineers, 1st Edition* (CRC Press)
[3] Dunning S and Katz L S 2017 *Energy Calculations and Problem Solving Sourcebook: A Practical Guide for the Certified Energy Manager Exam. 1st Edition* (Fairmont Press)
[4] Porpatham E Ramesh A and Nagalingam B 2012 *Fuel* **95** pp 247–256
[5] Porpatham E Ramesh A and Nagalingam B 2008 *Fuel* **87** pp 1651 – 1659
[6] Przybyła G Postrzednik S and Żmudka Z 2013 *Scientific Magazine - Combustion Engines* **3 (154)** pp 114-115
[7] Schroeder A Łach J and Poskrobko S 2009 *Archives of Waste Management* **11 (3)** pp 63–74
[8] Prauzner T and Ptak P 2011 *Journal of Technology and Information Education* **1** pp 292–296
[9] Herold K E Radermacher R and Klein S A 2016 *Absorption Chillers and Heat Pumps. 2nd Edition* (CRC Press)
[10] Pytel K Suchanicz J Livinsh M and Sternberg A 2013 *Condens. Matter Phys.* **16(3)** 31706
[11] Xu P Ma X Zhao X and Fancey K 2017 *Applied Energy* **203** pp 761–777
[12] Zhukov V Melnik O Logunov N and Chernyi S 2019 E3S Web of Conferences. **135** 02015
[13] Alpatov V and Soloviev A 2018 MATEC Web of Conferences **196** 01008

Acknowledgments

The project is financed by the Polish National Agency for Academic Exchange, No. PPI/APM/2018/1/00049/U/001.

IOP Conf. Series: Materials Science and Engineering 1016 (2021) 012022 doi:10.1088/1757-899X/1016/1/012022

Simulation and Analysis of the Interdigital Transducers Surface Acoustic Waves Generation and Propagation in Acoustophoretic Lab-Chip

O Matviykiv*, T Klymkovych, N Bokla and V Stakhiv

Department of Computer-Aided Design Systems,
Lviv Polytechnic National University, Lviv, Ukraine

*Email: matviykiv@gmail.com

Abstract. This work is devoted to the application of interdigital transducers for the formation of a standing acoustic wave in the microchannel of an acoustophoretic Lab-chip. 2D and 3D simulation of IDT operation was performed. The influence of IDT shape, its size, and location, geometric dimensions of the piezo-crystalline layer, as well as dimensions and material of the microchannel walls for the effective formation of standing acoustic waves were analyzed.

1. Introduction

The design, manufacturing, and application of microacoustofluidic Lab-chip devices combine technologies from solid and fluid mechanics, microelectronics, chemistry, biology and other domains [1-5]. Due to the processing of small volumes of liquids (from a few picolitres to several tens of microliters) the fluid flows in microfluidic devices comply with laminar flow (low Reynolds numbers), which allows to develop and use of various deterministic mechanisms for flow control. However, micro-and nanoscales require the consideration of surface-related phenomena like capillary forces, surface tension, roughness, fluid-structure interactions and others, which makes the Lab-chip engineering a complex multidisciplinary process. This is especially true for biological and biochemical mixtures, where non-contact transportation of microparticles is of crucial importance for the study of many biochemical phenomena. Typical non-contact methods based on electromagnetic principles have significant potential and are widely used in material science, but are limited for microfluidic Lab-chip applications because of particle size, fluid flow, certain material properties and/or other requirements. In the contrast, acoustic levitation is a non-contact and material-independent method of exploring microparticles, as well as does not require time-consuming sample preparation procedures.

The use of acoustic waves for focusing, separating, or transportation microparticles in the microchannels of Lab-chip devices attracts considerable attention from researchers due to the simple implementation of acoustic wave generators based on piezoelectric transducers. Since the sizes of acoustic waves generated by piezoacoustic transducers correspond to the characteristic sizes of microchannels, there are plenty of successful experiments with both surface and bulk acoustic waves. The design, manufacture and control techniques of such interdigital transducers (IDT) are well developed in microelectronics, but their direct use in microfluidic devices requires adaptation and additional research. Mathematical modeling of the acoustic radiation spreading as well as modeling of microparticles movement in acoustofluidic Lab-chip devices is still a design challenge, which should be overcome during the development of new sensors, studying new materials or analyzing wave propagation in new designs.

The coupling and accurate interpretation of complex microacoustofluidic mechanisms including the IDT SAW generation, wave propagation in Lab-chip design, leaking SAW in fluids and its interaction with dispersed microparticles are the main topics of further discussion. The present continues our previous research on the acoustophoretic motion of plastic microparticles in suspension [3].

2. The specifics of acoustophoretic Lab-chip devices

All acoustophoretic Lab-chip devices are based on an acoustic field, which combines two types of acting forces – *acoustic radiation force* F_{irad}, and the indirect *streaming drag force* F_{idrag} caused by acoustically induced, circulating, streaming flow rolls (Figure.1). The figure shows that standing acoustic wave formed in half-wave microchannel cause the acoustic force, which eventually focuses particles towards the low-pressure nodal plane due to second-order effects.

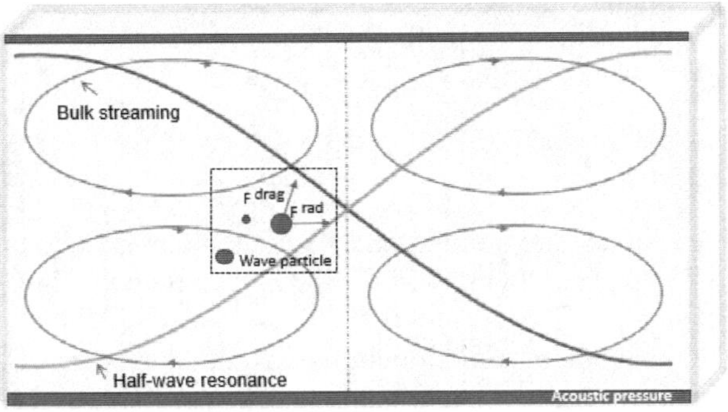

Figure 1. Schematic view of microchannel cross-section represented acoustic forces and streaming flow rolls and forces acting on microparticles.

Depending on the design of the piezoelectric transducer, Lab-chip devices can utilize bulk acoustic waves (BAW) or surface acoustic waves (SAW) [6]. BAW devices are actuated through electrodes attached to both sides of a piezoelectric substrate, which generates acoustic waves in the entire transducer body. In SAW devices actuation is achieved through a pair of interdigitated opposite charged electrodes (IDT) formed on the piezoelectric substrate or layer. Further research here will be focused on SAW devices which are based on the Rayleigh mode. These waves include both longitudinal and transverse motions transmitted along the surface of the piezoelectric substrate and cause subnanometer movement of its molecules. At the point where SAW interacts with the liquid it changes its vibration mode into a leaky longitudinal wave which continuously travels through the liquid at a Rayleigh angle [7]:

$$\theta_R = sin^{-1}\frac{v_L}{v_S},$$

where v_S – SAW velocity on the piezoelectric substrate and v_L – acoustic velocity in the liquid.

Using two pairs of orthogonally opposing IDTs that generate opposite leaky waves, one can obtain in microchannel or chamber the standing acoustic wave at ultrasonic frequencies with formed acoustic nodes and antinodes. These pressure nodes form streamlines in the microchannels and focus suspended particles similar to hydrodynamic or electrokinetic effects.

3. Analysis of piezoelectric SAW generators for Lab-chip devices

A typical acoustofluidic SAW Lab-chip device consists of a substrate covered by a thick piezoelectric lithium niobate layer with one or two aluminum IDT placed at the opposite sides of the rectangular microfluidic channel (Figure2). By applying an oscillating electrical signal to IDT contacts, we activate SAW at a resonant frequency [8]:

$$f = \frac{c_S}{\lambda_{SAW}},$$

where c_S is the sound speed in lithium niobate, and λ_{SAW} is the spacing between successive IDT finger-pairs.

In general, for defined piezoelectric material (known speed of sound), the resulting SAW frequency, its amplitude, and wave-front orientation are defined by the design and dimensions of the IDT electrodes, frequency and input power of the applied electrical signal.

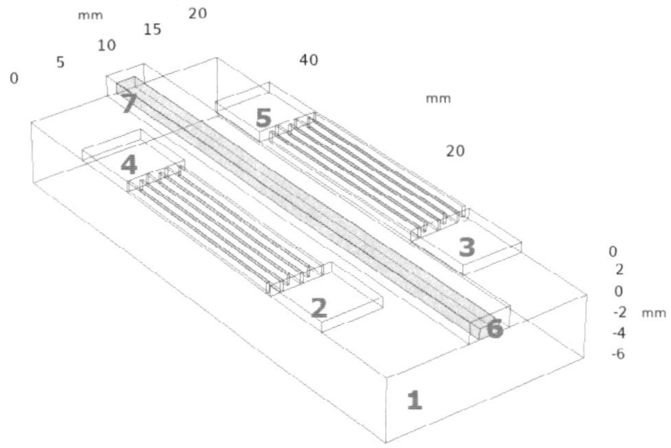

Figure 2. Schematic design of acoustofluidic Lab-chip.
1 - a piezoelectric layer of LiNbO3; 2-5 - IDTs, 7 – microchannel of PDMS, 6 - liquid.

However, depending upon required functionality there exist several IDT configurations, which have recently been used in various Lab-chip applications. The main design is presented in Figure 3.

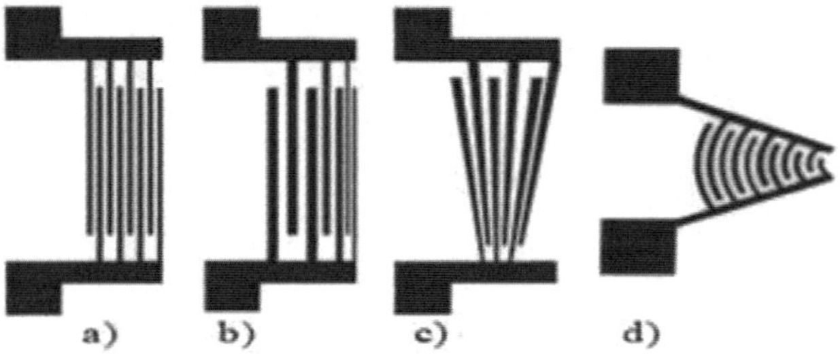

Figure 3. Typical IDT configurations in various Lab-chip designs:
a) Straight IDT; b) Chirped IDT; c) Slanted IDT; d) Focused IDT [9].

The straight IDT design (a) is a fundamental one, which consists of uniform electrode fingers, where each metal strip has a width of ¼ λ_{SAW}. In specific cases of narrow microchannels when lateral displacement needs to be formed, a concentric circular arc structure of focused IDT (d) may be used for generating SAW with higher intensity. Next transducer configurations, like Chirped IDT (b) and slanted IDT (c), respectively, consist of fingers with varying spacing and dimensions to enable operation at a range of resonant frequencies [10].

4. Simulation of IDT for Lab-chip devices

The Lab-chip device analyzed in this work consists of a rectangular microfluidic channel with a cross-section 1,0x1,0 (mm) made by PDMS material. SAWs were generated by straight IDT designed with sets of 2, 3, 4 and 5 connected metallic finger pairs. The IDT fingers with 12,0 mm length, 10,0 mm aperture (overlapping width) and 0,25 mm pitch were formed by 1,0 μm aluminum film on top of the piezoelectric layer. In this simplest equidistant IDT, which has a constant spatial half-period of the electrodes $h_n = \lambda_0 / 2$ and a constant electrodes length, the partial waves are added in phase with the same amplitude. As a result, the frequency response of such transducer equal to the IDT acoustic synchronism frequency has the form of $\sin x / x$. To increase the selectivity of the IDT and implement various frequency characteristics, a variety of IDT designs have been proposed for weight processing of the amplitudes and phases of partial waves.

IOP Conf. Series: Materials Science and Engineering 1016 (2021) 012022 doi:10.1088/1757-899X/1016/1/012022

3D simulation of IDTs in the COMSOL environment allows obtaining the electrical potential on the surface of the $LiNbO_3$ electric potential (V) when applying an RF signal with an operation frequency of 1,48 MHz to one contact pad, while the other is grounded (Figure 4). The simulation results show that the penetration depth of the electrical signal is about 4.0 mm.

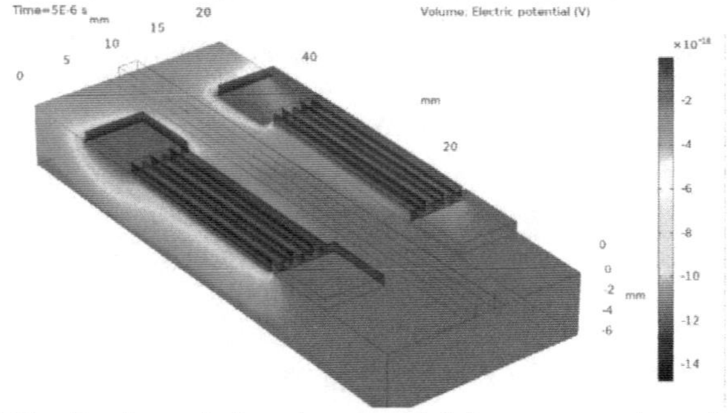

Figure 4. Distribution of electric potential for acoustophoretic Lab-chip.

To analyze the influence of various SAW lengths (channel width 2λ, λ and λ/2) and microchannel location on the formation of stable standing acoustic waves, several simulations were carried out (Figure 5).

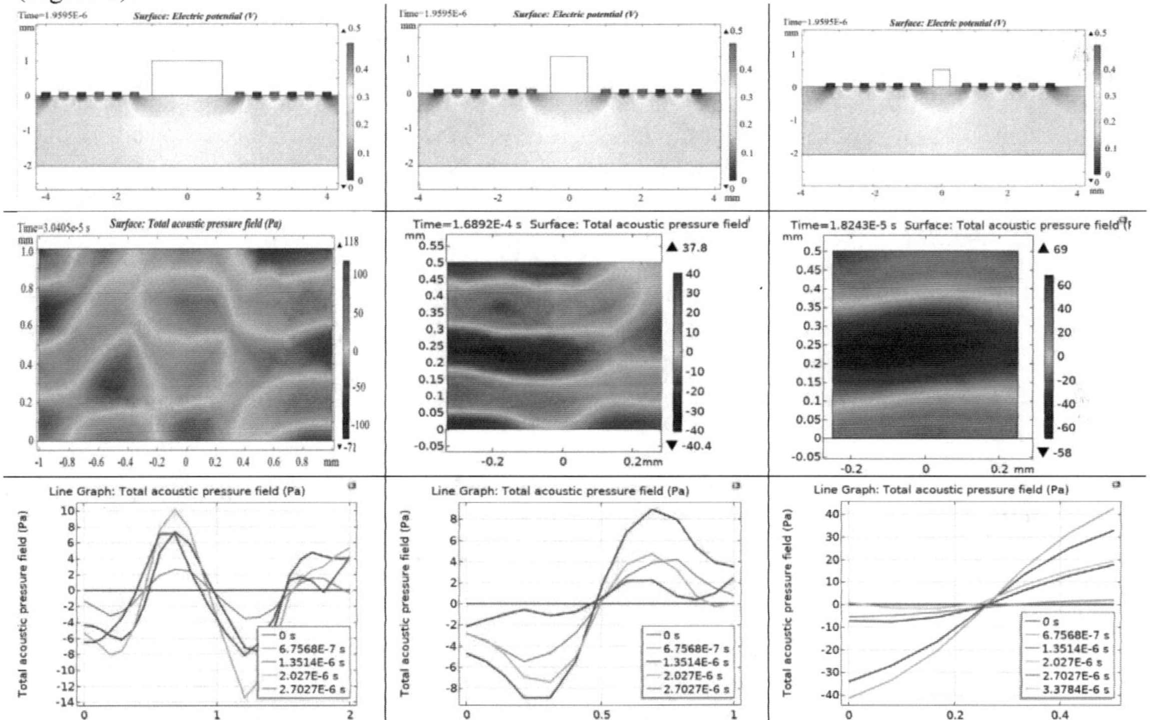

a) $w_{ch}= 2\lambda$ b) $w_{ch}= \lambda$ c) $w_{ch}= \lambda/2$

Figure 5. Simulated surface electric potential (upper row), surface acoustic pressure field (middle row), and contours of standing acoustic wave in microchannels with 2λ (a), λ (b), and λ/2 (c) channel width.

The top row of the figure shows the distribution of the electric potential for a three-finger IDT. The excitation RF source generated a sinusoidal wave *(sin (2π / T_0 t))*. The maximum value of the potential is 0.37 V for the channel's width of *2λ* and *λ* and 0.36 V for the channel width of *λ/2*. As can be seen from all three figures, the total penetration depth of the electric potential under IDT equals *λ/2*, and at

λ depth, the electric potential is close to zero. So, in practice, the thickness of the piezoelectric layer needs to be at least one λ or 1.0 mm in our case.

The middle row of the figure shows the distribution of the acoustic pressure field. The acoustic pressure varies between -3,76..2,48 Pa for the channel width of 2λ; -5,36..3,75 Pa for the channel width λ; and -5,89..4,14 Pa for the channel width $\lambda/2$. The positive pressure value is shown in red, and the negative – in blue. The area depicted in white means zero pressure regions, where acoustic standing wave nodes are placed.

The focuses of the standing wave are well shown in the bottom row of the figure. It presents linear graphs of acoustic pressure along the line parallel to the channel bottom at a height of $\lambda/4$. Each line corresponds to a time step from 0 to 6,7568E-6 s with an interval equal to the wave period $T_0 = 1/f_0$, where f_0 is the frequency of the wave. As can be seen from Figure 5, at channel width 2λ three acoustic focuses appeared, at channel width λ or $\lambda/2$ – one acoustic focus. However, for the channel width $\lambda/2$, the acoustic focus is more accurate. Considering that for channel width $\lambda/2$ the acoustic pressure gradient is higher, one can assume that for such condition the focusing of suspended particles will be the most stable.

The theoretical analysis and obtained simulation results confirmed that IDT's operational efficiency is determined by the substrate and electrode materials, IDT electrodes aperture, width, and height. These factors can be easily optimized by certain design parameters, like operation frequency. Additionally, the number of finger pairs can be modified to adjust the IDT impedance of the excitation, which plays a critical role in achieving efficient operation of SAW.

Conclusions

The increased popularity of the SAW-driven Lab-chip devices requires a fundamental understanding of physical phenomena appeared in IDTs, piezoelectric substrate, microchannel walls, fluid mixture and solid particles in fluid flow. However, the effective generation of SAW in acoustofluidic Lab-chip devices can be possible only through the design and optimization of the IDTs structure with the help of the CAD/CAE tool.

To increase the performance and stability of IDT, it is important to reduce signal loss and avoid unwanted side effects. It is shown that in addition to the choice of the piezoelectric substrate material with a high coefficient of electromechanical coupling, this can be achieved by optimizing the design and parameters of IDT, as well as optimizing the size and location of the microfluidic channel:

1. For the efficient manipulations with microparticles dissolved in the fluid flow, the optimal microchannel configurations have a width from $\lambda / 2$ to 2λ and a height equal to either the wavelength λ or twice less $\lambda /2$.
2. The thickness of the piezoelectric layer (lithium niobate) critically affects the formation of standing acoustic waves only when it was reduced up to $\lambda / 2$. To prevent such edge effects which lead to a decrease in the acoustic pressure in the microchannel, the thickness of the piezoelectric layer should be greater than 2λ.
3. The number of IDT fingers has an important influence on the time of standing acoustic wave stabilization. For the studied configurations, the 4-finger IDT was the most optimal. Calculations proved that for such structures, the stable distribution of acoustic pressure in the microchannel was achieved in 10 cycles, and the standing acoustic wave is fully formed within 0.12 s.

References

[1] Matviykiv O Klymkovych T Farmaha I Bokla N Marikutsa U and Pytel K 2020 Thermal Design of Microreactor Structural Elements for Water Quality Analyser *Journal of Physics: Conference Series* **012007** doi: 10.1088/1742-6596/1534/1/012007

[2] Lobur M Vivchar D and Jaworski N 2016 MEMS pressure sensors design by the COMSOL system *Modern Problems of Radio Engineering, Telecommunications and Computer Science, Proceedings of the 13th International Conference on TCSET* 156-158 doi: 10.1109/TCSET.2016.7451997

[3] Matviykiv O Klymkovych T Bokla N Lobur M Melnyk M and Timofiejczuk A 2020 Simulation of Acoustophoretic Separation of Microplastic Particles in mkFluidic Lab-chip *Proc. Int. Conf. on the Perspective Technologies and Methods in MEMS Design (MEMSTECH)* 123- 6

[4] Klymkovych T Bokla N and Matviykiv O 2019 Microfluidic Lab-Chip Device Dedicated for Colorimetric Detection of Hazardous Impurities in Water Samples *IEEE XVth International Conference on the Perspective Technologies and Methods in MEMS Design (MEMSTECH)* 96-100 doi: 10.1109/MEMSTECH.2019.8817402

[5] Kosobutskyy P Jaworski N Farmaha I and Kuzmynykh M 2019 Optimization of probe parameters of atomic force microscope cantilever *IEEE XVth International Conference on the Perspective Technologies and Methods in MEMS Design (MEMSTECH)* 127-130 doi: 10.1109/MEMSTECH.2019.8817389

[6] Winkler A Bergelt P Hillemann L and Menzel S 2016 Influence of viscosity in fluid atomization with surface acoustic waves *Open Journal of Acoustics* **vol 6(03)** 23-33

[7] Weser R Winkler A Weihnacht M Menzel S and Schmidt H 2020 The complexity of surface acoustic wave fields used for microfluidic applications *Ultrasonics* **106**

[8] Fu Y Luo J Nguyen N et al. 2017 Advances in piezoelectric thin films for acoustic biosensors, acoustofluidics and lab-on-chip applications *Progress in Materials Science* **89** 31–91

[9] Fakhfouri A 2019 The design of surface acoustic waves field for manipulation of particles in microfluidics *Monash University. Thesis*

[10] William B Lei J Thomas L and Henrik B 2020 Microparticle Acoustophoresis in Aluminum-Based Acoustofluidic Devices with PDMS Covers *Micromachines* **11**(3) 292

IOP Conf. Series: Materials Science and Engineering 1016 (2021) 012023 doi:10.1088/1757-899X/1016/1/012023

Experimental Study of the Time of Reverberation of a Sacred Building Having a Cruciform Architecture

M Melnyk[1]*, A Kernyskyy[1], Y Mykhalyna[1] and K Pytel[2]

[1] Lviv Polytechnic National University, Lviv, Ukraine
[2] AGH University of Science and Technology, Krakow, Poland

*Email: mykhaylo.r.melnyk@lpnu.ua

Abstract. The response of the interior to the excitation of the pulse noise was registered using the Svan958A spectrum noise analyzer. After that, using the Dirac system, the registered files were analyzed and the time of reverberation of the church of St. John the Baptist, located in the village Verkhnya, Kalush district, Ivano-Frankivsk region, Ukraine with indication of measuring points. Detailed measurements of the investigated object were carried out and on their basis a 3D-model of the church in the SketchUP system was developed, which allowed to determine the recommended reverberation time for this type of interior and the given volume. The conducted researches made it possible to determine the uniformity of sound energy distribution in building having a cruciform architecture.

1. Introduction

The main acoustic parameters used to assess the acoustic quality of sacred interiors for many years one of has been the reverberation time [1-2]. The calculated or measured value of the reverberation time can be compared with the values recommended for this type of interior and thus assess whether there will be sufficient language intelligibility and volume in all places where believers may be [3]. Reverberation time is still the main and most frequently used factor that determines the acoustic properties of the sacred interior due to the fact that there is still no single method of assessing the acoustics of this type of interior, and because it is an internal parameter that can be calculated at the design stage and then tested experimentally [4-6].

The evaluation based on the comparison of the measured reverberation time to the recommended one does not fully reflect the acoustic qualities of the interior of a church designed to perform many functions in terms of acoustics. However, this parameter allows an introductory assessment of the ability of the assessed premises to perform its function in terms of acoustics [7].

That is why the task was set to investigate the acoustic properties of the St. John the Baptist church, located in the village Verkhnya, Kalush district, Ivano-Frankivsk region (see Fig. 1).

For this purpose, it is necessary to carry out measurement of the geometrical sizes of a construction along with materials. The next step is to conduct an experimental study of the reverberation time in order to find the uniformity of the distribution of sound energy in the premise of study. The source of impulse noise must be installed at the place where the priest usually conducts the service, and the second option is to install the source where the church choir is located.

Figure 1. Photo of the object of study.

2. Construction of a 3D model of the church

In the literature, the recommended reverberation time is very often tied to the volume of the interior [7]. To accurately determine the volume of the interior, it was proposed to develop a 3D model in the SketchUP system. The developed model made it possible to determine precisely the volume and this can also be used to build an acoustic model of the church when exporting to room acoustics modeling systems, such as CattAcoustic in order to further acoustically adapt the church premises [8]. The model can also be used to test the accuracy of acoustic modeling systems, but it is necessary to specify the absorption coefficients used in the materials of the church decoration. Most often, the model has discrepancies with experimental data due to inappropriate setting of sound absorption coefficients of materials.

At the initial stage, it was proposed to develop a model according to the drawings (see Fig. 2.), but the test showed that the actual dimensions do not always correspond to the drawing, so using the laser rangefinder the actual dimensions were taken.

Figure 2. Church drawing (side view).

A 3D model of the church was designed according to the actual dimensions, which is presented in Fig. 3. The church is made of brick and covers an area of 397 m^2 with a volume of 4070 m^3. The height of the dome is 22.5 m.

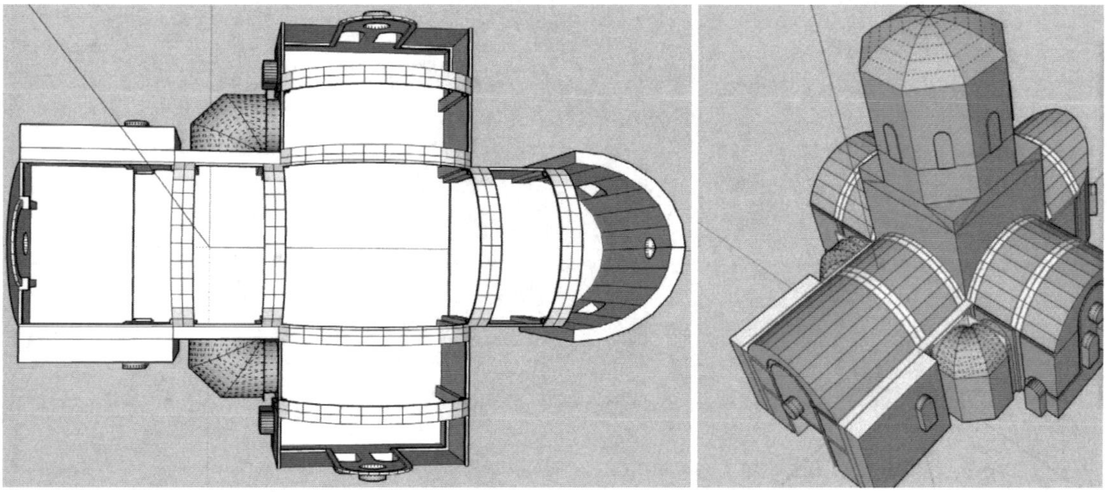

Figure 3. 3D model of the church

3. Description of measurement conditions

The signal for the tests and experiments was an impulse noise. Balloons are used as a source of pulsed noise, which fully meet the requirements for pulsed noise sources. The source of pulsed noise was installed at a height of 1.5m and 1.2m from the floor [9]. A capacitive omnidirectional measuring microphone SV22, set at a height of 1.2m, and a noise meter spectrum analyzer SVAN 958A were used to record the response of the interior [10, 11]. The equipment was calibrated before the experiment. All experiments were performed at a temperature of 19 °C and a relative humidity of 49%. The recorded signals were processed using an algorithm implemented in the Dirac 5.0 system. As a result of its work, the reverberation time was determined in the octave ranges from 125 to 8000 Hz. The photo of the experiment is presented in Fig. 4.

Figure 4. Reverberation time measurement.

4. Comparison of the results of the experiment with the recommended values of the parameters

To compare the values of the acoustic parameters obtained as a result of the experiment with the recommended values, the average values of T30 (T30 – the reverberation time for 60 dB of room decay, based on a straight line curve fit between the -5 dB to -35 dB points on the Schroeder curve [9]) for 500 and 1000 Hz were determined (see Table 1.).

Table 1. Statistical results T30 for 500 and 1000 Hz.

Frequency	500 Hz	1000 Hz
Number of experiments	11	11
Minimum	4,007	3,741
Maximum	4,244	4,029
Average value	4,142	3,91
Recommended value	3,54	3,54

The recommended value of the reverberation time for Catholic churches according to Kulowski [3] with a volume of 4070 $м^3$ marked as the average time from the graphical ratios T(V) is 1.82 sec. Where T is the reverberation time and V is the room volume. As we can see, the church does not meet the requirements, but it is difficult to clearly assess this due to the large discrepancy between the recommended reverberation time for churches. The tolerance range is ± 20%, except for the frequency 125 Hz (50% of the tolerance up) and 4000 Hz (40% of the tolerance down).

In order to investigate how the reverberation time changes depending on the observation point, as well as near the walls, in the middle and on the balcony, several measuring points were selected. The layout of the points where the measurements were carried out is presented in Fig. 5

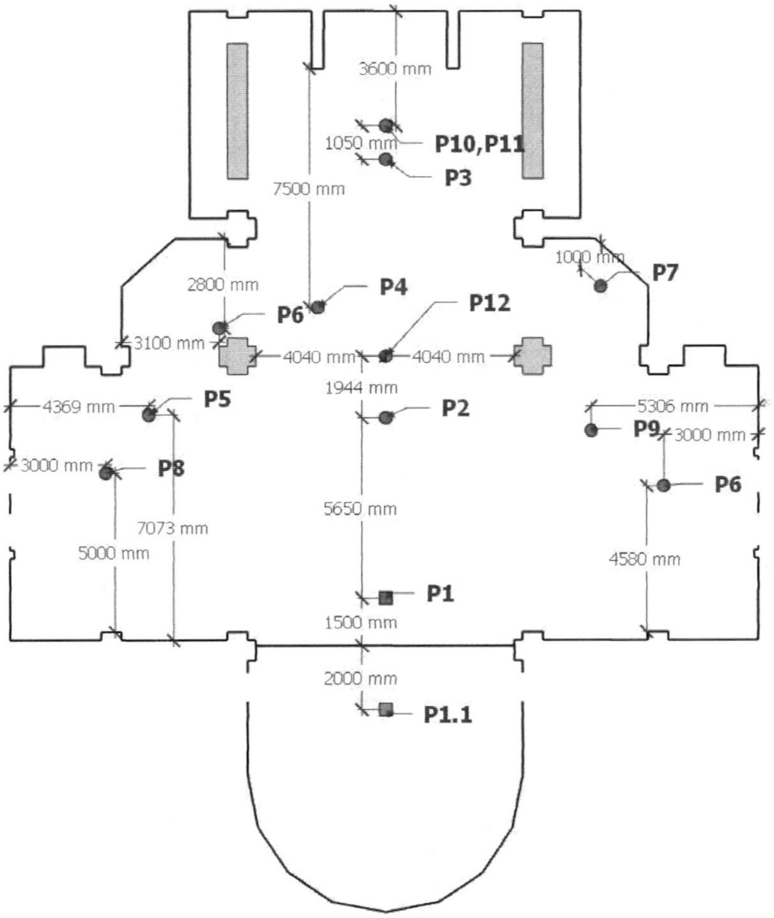

Figure 5. Layout of measuring points.

The graph (Fig. 6.) shows that the recurrence is at a high level. Only at low frequencies does point 3 differ from all the others, and this is due to the fact that it was almost under the balcony and the delayed sound energy coming from under the balcony was added, because it is the low frequencies that are poorly attenuated in the air.

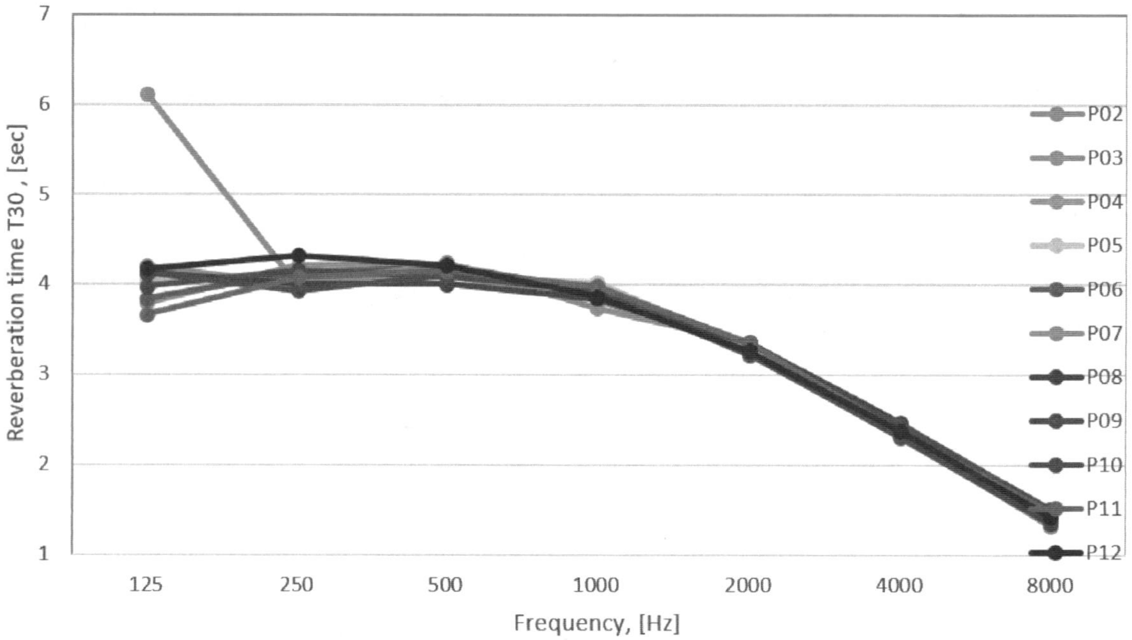

Figure 6. The results of determining the reverberation time for 11 different locations of the noise source and noise meter.

As mentioned earlier, for a church space of 4070 meters^3, the recommended reverberation time should be 1.82 seconds, and in our case, the average values for 500 and 1000 Hz are 4.1 seconds and 3.9 seconds, respectively. The first measurements were made without people because they would create additional noise. If we take into account the absorption surface of people and carpet, which was taken from the floor, we can assume that the acoustics of this church is rather good, but to confirm this fact it will be necessary to conduct additional simulations and use these results later to calibrate the model.

Conclusions

Studies of the reverberation time of the church have shown that the problem of sacred objects is their large volume, which in turn leads to an increase of the reverberation time. This fact is useful since the choir sings in churches most of the time, but it comes to the preaches, the intelligibility of the priest language is rather poor. Experiments have shown that the room distribution of echo time is uniform, but it is too large and the quality of speech transmission is therefore low.

References

[1] Beranek L L and Hidaka T 1998 Sound absorption in concert halls by seats, occupied and unoccupied, and by the hall's interior surfaces *The Journal of the Acoustical Society of America*, **104**(6), 3169-77.
[2] Engel Z et al 2007 *Podstawy akustyki obiektów sakralnych* ITE Kraków
[3] Kulowski A 2011 *Akustyka sal-Zalecenia projektowe dla architektów* Politechnika Gdańska
[4] Rubacha J et al 2019 Analysis of the Acoustic Parameters of the Maria Zankovetska Theatre in the Lviv Before and After Modernisation of the Audience *MS&E* 471(8) 082035
[5] Kosobutskyy P and Ferens R 2017 Statistical analysis of noise measurement system based on accelerometer-gyroscope GY-521 and Arduino platform *14th International Conference The Experience of Designing and Application of CAD Systems in Microelectronics (CADSM)* 405-7 doi: 10.1109/CADSM.2017.7916162
[6] Melnyk M Lobur M Vasyliuk I Mazur V and Hemich N 2011 The mathematical model of height position calculation of dominated noise source in a traffic flow *11th International*

IOP Conf. Series: Materials Science and Engineering 1016 (2021) 012023 doi:10.1088/1757-899X/1016/1/012023

Conference The Experience of Designing and Application of CAD Systems in Microelectronics (CADSM) 429-30

[7] Rettinger M 1977 *Acoustic Design and Noise Control: Noise control Vol 2* Chemical Publishing Company

[8] Melnyk Mykhaylo et al 2018 SketchUP to Catt-Acoustic Converter of Interior Design 3D Models *In: 2018 XXIIIrd International Seminar/Workshop on Direct and Inverse Problems of Electromagnetic and Acoustic Wave Theory DIPED* 165-9

[9] Standart ISO 3382-1 2009 - Acoustics -- Measurement of room acoustic parameters -- Part 1: Performance spaces.

[10] Svantek Instrumentation for sound and vibration measurements SVAN PC+ software User Manual ver. 2.1, SVAN 958A Four Channels Sound and Vibration Analyser 2015

[11] Pytel K et al 2019 The Creativity of the Application of MEMS in the Implementation of Technical Activities as a Result of Engineering Education in the Field of Environmental Protection *IEEE XVth International Conference on the Perspective Technologies and Methods in MEMS Design (MEMSTECH)* 157-160

Acknowledgments

The project is financed by the Polish National Agency for Academic Exchange, No. PPI/APM/2018/1/00049/U/001.

AUTHOR INDEX